T0197932

Seaweeds of the British Isles

Seaweeds of the British Isles

A collaborative project of the British Phycological Society and the Natural History Museum with financial support from the Joint Nature Conservation Committee

Volume 2

Chlorophyta

Elsie M. Burrows

Natural History Museum, London

First published by the Natural History Museum,
Cromwell Road, London SW7 5BD
© Natural History Museum, London, 1991

This edition printed and published by Pelagic Publishing, 2013,
in association with the Natural History Museum, London

The Authors have asserted their right to be identified as the Authors of this
work under the Copyright, Designs and Patents Act 1988.

ISBN 978-1-907807-72-5

This book is a reprint edition of 1-898298-85-8.

All rights reserved. No part of this publication may be transmitted in any
form or by any means without prior permission from the British Publisher.

A catalogue record for this book is available from the British Library.

Seaweeds of the British Isles
Publishing in seven volumes

* Part 1 Introduction, Nemaliales, Gigartinales.
P.S. Dixon & L.M. Irvine 1977.
Part 2A Cryptonemiales (*sensu stricto*), Palmariales,
Rhodymeniales. L.M. Irvine 1983.
** Part 1 R.L. Fletcher 1987.
† T. Christensen 1987.

Contents

Acknowledgements

During the course of preparation of this volume, I was helped by very many people with specimens, literature, discussions and encouragement and would like to express my gratitude to all of them. In particular I would like to thank:

Professor P. S. Dixon who did a great deal of work on the nomenclature and provided me with help in many ways, Professor F. E. Round who gave me much help and encouragement in finishing it, Dr Jane Lenton who inked in and prepared my original pencil drawings for publication, Dr G. Boalch who continually kept me in touch with recent work, the late Professor M. de Valera for her help and hospitality on many occasions; also Dr Y. Butler, Dr W. F. Farnham, Dr R. L. Fletcher, Dr D. E. G. Irvine, Dr C. Maggs, Dr G. Russell and the late Dr M. Parke.

I am grateful to the algal research workers in the Botany Department of the University of Liverpool who contributed so much to this volume and to whose work reference is made in the text; to my collaborators Dr M. Wilkinson, who made the original drafts for the descriptions of *Codiolum polyrhizum*, *Eugomontia sacculata*, *Entocladia tenuis*, *Monostroma grevillei*, *Tellamia contorta* and *Tellamia intricata*, and Professor G. R. South who drafted the description for *Acrochaete repens*.

I am indebted to the Liverpool University Joint Committee on Research for a travel grant in the early stages of the work; also to the Directors and Curators of the following institutions for working and laboratory facilities and permission to examine and borrow specimens:

Department of Botany, British Museum (Natural History)
Royal Botanic Gardens, Kew
The Linnean Society of London
Marine Biological Station, Port Erin, Isle of Man
Furzebrook Research Station, Dorset (now The Institute of Terrestrial Ecology)
Department of Biology, Memorial University of Newfoundland

The Laboratory of the Marine Biological Association of the United Kingdom at Plymouth

The Nova Scotia Research Foundation, Seaweeds Division, Halifax, Nova Scotia

The Lough Ine Marine Laboratory of the late Professor J. A. Kitching to whom I owe special thanks.

I am also grateful to the British Museum (Natural History) for a grant towards the production of the illustrations.

Foreword

Anyone contemplating writing a flora should not underestimate the formidable task they have set themselves, beginning with the painstaking evaluation of all the species records through to the final writing of the individual entries and preparation of the keys and figures. In agreeing to write the volume dealing with the green algae for the Seaweeds of the British Isles, Dr Elsie Burrows had no illusions as to its magnitude but could not have forseen that its publication would not be in her lifetime.

Dr Burrows (Bunny to her friends) began compiling information for the Flora as long ago as 1951 and continued its preparation following her retirement to Dorset in 1973. Although drawing heavily upon her own considerable experience, she was always willing to seek the opinions of other specialists and to involve her students in researching problems relating to individual taxa. We know she was very grateful to everyone who helped her in this and many other ways and we would like to record our thanks to them on her behalf. After her collaborator, Sheila Lodge, resigned in 1955, Dr Burrows always made it known that she was solely responsible for seeing the 'Greens' through to completion and was unwilling to relinquish the task despite failing health. A draft manuscript was completed shortly before her death in September 1987. Unfortunately she died before she was able to undertake the meticulous checking of the draft and it was then left to the Flora Committee of the British Phycological Society to arrange for the posthumous preparation of the text and figures for publication. In accordance with her wishes only minor adjustments have been made to her original manuscript, mostly taking the form of correcting orthographic errors or dealing with omissions. The Flora Committee wish to record their appreciation to all those who gave so generously of their time in order to ensure the publication of Dr Burrows' manuscript. Special mention and thanks go to the following: Professor P. S. Dixon, Dr R. L. Fletcher, Mrs L. M. Irvine, Mr P. W. James, Dr D. M. John, Professor F. E. Round and Mr I. Tittley. The Committee also wishes to thank Dr Burrows' daughter, Mrs Jill Rodgers, for giving us access to literature and specimens after Dr Burrows' death.

The Flora Committee believed it important that the author's last wishes be respected, and trust this, her final work, will serve as a lasting tribute to the memory of a much respected member of the phycological community. Her obituary by Professor T. A. Norton (1987) outlines the important role she played in British post-war phycology.

We thank her for the tremendous effort she put into writing this volume, often under very difficult circumstances, and only wish she could have been alive today to see the successful outcome of her labours.

The Flora Committee,
British Phycological Society

Abbreviations

The following standard abbreviations are used for herbaria cited in the text.

AMD   Hugo de Vries Laboratories, Hortus Botanicus, Amsterdam, Netherlands.
BM   British Museum (Natural History), London.
BM-K   The algae previously at the Royal Botanic Gardens, Kew, now on permanent loan to the BM.
BORD   Jardin Botanique de la ville de Bordeaux, Bordeaux, France.
C   Botanical Museum and Herbarium, Copenhagen, Denmark.
CN   Laboratoire de Botanique, Faculté des Sciences, Caen, France.
CO   Laboratoire de Biologie Marine, Concarneau, France.
HELG   Biologische Anstalt, Helgoland, Germany.
KIEL   Botanisches Institut der Universitat, Kiel, Germany.
L   Rijksherbarium, Leiden, Netherlands.
LD   Botanical Museum, Lund, Sweden.
LINN   Linnean Society, London.
O   Botanisk Museum, Oslo, Norway.
OXF   Fielding – Druce Herbaria, The University, Oxford.
PC   Laboratoire de Cryptogamie, Muséum National d'Histoire Naturelle, Paris, France.
RO   Istituto Botanico, Città Universitaria, Rome, Italy.
S   Naturhistoriska Riksmuseum, Stockholm, Sweden.
TCD   Department of Botany, Trinity College, Dublin.
UC   Herbarium, University of California, Berkeley, California, USA.
WU   Naturhistoriches Museum, Vienna, Austria.

Introduction

The objective of this volume is to provide a means by which workers can identify marine, green algae collected from the field. For this purpose, a morphological concept of the algal species is used, i.e. in the definition of Davis & Heywood (1963) it is 'an assemblage of individuals with morphological features in common and separable from other such assemblages by correlated morphological discontinuities in a number of features'. Specimens collected from the field are samples of populations and species delimitation depends on the presence of breaks in the morphological variation shown by these populations. The population in this sense means 'an assemblage of plants with a particular distribution; its parameters are judged from morphological evidence' (Davis & Heywood, 1963). It has not been found possible to include an interfertility/intersterility criterion in the species concept, such as has been employed for *Enteromorpha* species by Bliding (1963) because firstly, few data are available for the marine green algae and secondly, identification has often to be made on non-fertile material and most importantly because only positive data are of any use since the production of viable gametes and their fusion for cross-breeding experiments depends so much on the condition of the plants when collected from the field and the particular laboratory conditions used.

The classification used in this volume is essentially phenetic, based as it is on morphological and easily determined structural characters and reproductive bodies. It is essentially the system used in the Check-List of British Marine Algae (Parke & Dixon, 1976), but altered to some extent for reasons discussed at the appropriate places in the text. The phenetic character of the classification may not reflect relationships to the degree which early workers aspired: only by combining the morphological and life history data with that from biochemical and ultrastructural studies will an overall phylogenetic system emerge. The new ideas, derived especially from ultrastructural studies of cell division and of reproductive cells and biochemical studies of cell wall and other constituents have led to suggestions for major new systems of classification which differ fundamentally from those held until recently. A detailed account of the new systems summarizing most of the work concerned can be found in the Systematic Association's special volume No. 27, 1985 (edited by Irvine & John). Discussion of such classifications is still in a fluid state and no system is yet generally accepted. As has been pointed out by van den Hoek in Lobban & Wynne (1981) the new concepts are based almost exclusively on ultrastructural characters and these are only a fraction of the characters which must be used to classify taxa. A final, natural classification needs to be based on all available characters. It may be desirable to have a phenetic classification running side by side with a phylogenetic system, the two serving different purposes. To that extent the system used in this volume may be regarded as artificial, though it is possibly no less so than classifications based largely on biochemistry or ultrastructure.

No general discussion of the green algae is provided by this volume as there are many comprehensive accounts in the literature and the reader is referred to these, e.g. Fritsch (1935), Bold & Wynne (1978), Lobban & Wynne (1981). The present volume provides descriptions and keys for the identification of species of the benthic marine Chlorophyceae occurring around the British Isles. There are marine representatives of most of the orders in the Class, though for some, such as the Chlorococcales, the bulk of the taxa occur only in freshwater and yet other orders have no marine representatives. Purely planktonic algae, including all of the marine Volvocales and most of the Chlorococcales are omitted. A few unicellular species of the Chlorococcales have been included here because they occur in association with benthic species and some of them may possibly be stages in the life history of larger green marine algae.

Essentially freshwater benthic species which just spread into low salinity brackish water, e.g. *Cladophora glomerata* and *Cladophora fracta* have been omitted from the volume, but brackish water species have been included if they spread into more truly marine conditions, e.g. *Cladophora vagabunda*.

Distribution records for the coasts of the British Isles are quoted in terms of the counties system in use before 1974 on the basis of which records for the earlier volumes in the series have been given. The map of the counties in Volume I Part 2A (Rhodophyta) is reproduced here to aid the reader. In determining distributions in the British Isles, on the whole only records that have been checked by the author and by specialists in particular taxa are included. When older records are used an attempt has been made to check that the specimens recorded conform to the present interpretation of the species. There are inevitably big gaps in the distribution records, particularly for algae that are present only or mainly during the winter months. The fact that records are as complete as they are is a tribute to the work of members of the British Phycological Society at its field meetings and of individual workers at other times; many have sent in specimens for checking in relation to the Society's Mapping Scheme (Norton, 1985) and these have proved most valuable.

Seasonal behaviour, growth and reproductive periods of the species have been difficult to establish because, although records of the presence of a species may be common, the condition of the plants frequently have not been given. Much remains to be achieved on these aspects, which are perhaps best tackled through detailed autecological studies of individual species or genera.

Vertical distributions are described in relation to the shore zones – upper, middle and lower littoral, and sublittoral. The terminology is the same as that discussed and used in Volume I Part I (Rhodophyta) of this series and illustrated in Fig. 12 of that volume. The figure is based on Lewis (1964) and is reproduced here for quick reference (Fig. 1).

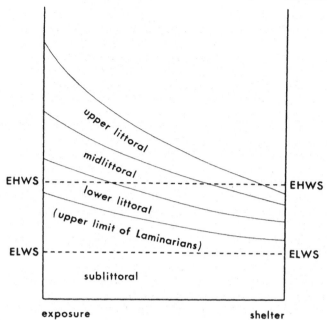

Fig. 1   The terminology for zonation used in the text. The position of the highest spring tide is indicated by EHWS; the position of the lowest spring tide is indicated by ELWS. Based on Lewis (1964).

The zones are defined in terms of both the movement of tides and the presence or absence of key organisms.

Littoral zone

The *upper littoral zone* is characterized by organisms such as species of *Littorina* and the lichen *Verrucaria*. It encompasses the upper range of the tidal cover on sheltered shores and extends above tidal limits only on shores exposed to strong wave action.

The *midlittoral* and *lower littoral* zones are characterized by barnacles, limpets and *Fucus* species and are subjected to daily alternating exposure of air and wetting by tidal immersion. The boundary between mid- and lower littoral depends on the shore and on convenience in different examples.

Sublittoral zone

The *sublittoral zone* is characterized by organisms which are totally or almost totally immersed with only the uppermost parts uncovered for short periods on each tide. The upper limit of the zone is at the point where *Laminaria* species cease to grow on rock surfaces.

The distribution of the green marine algae on the shores of the British Isles is essentially similar to that in northern Europe. There are few, if any, species that are confined to the British Isles. *Sykidion dyeri* Wright (1881), not recorded elsewhere in Europe, is of uncertain identity.

The green marine algae belong mainly to the littoral and shallow sublittoral zones of the shore and are subject to constantly changing and sometimes extreme environmental conditions, depending on weather. The British Isles lie in a zone where the mean annual surface temperature ranges between 5–15°C. Both winter and summer sea surface temperatures differ, not only between northern and southern shores but also between east and west, the western shores being warmer. However, as has been pointed out by Lewis (1964), the length of tidal exposure, degree of insolation and other changing factors such as rainfall, probably affect the growth of the algae more than the actual sea temperatures. In spite of the fact that the factors vary considerably round the coast, the majority of marine green algal species occur around the entire coastline if suitable habitats occur. It is to be expected, therefore, that adaptability is an important character for such algal species and this would appear to be the case.

One or two species could perhaps be regarded as 'northern' in that they do not reach the south coast of England, e.g. *Codium fragile* subsp. *atlanticum* which reaches Bardsey Island in the Irish Sea and the Farne Islands in the North Sea. The reason for this is not clear, but it would appear not to be a temperature barrier as it occurs all the way down the west coast and on to the south coast of Ireland. A few species of marine green algae belong to somewhat warmer waters and reach as far north as the south coast of England and the south west coast of Ireland, e.g. *Ulva olivascens*. Whether this is a recent invasion during very hot summers since 1976 or whether it was here earlier unobserved is uncertain, but it appears now to be frequent in Cornwall.

Another northern species is *Cladophora pygmaea* which until recently was known only from the north of Scotland. Now it is known from the sublittoral zone of the west coasts of Scotland and Ireland and has been recorded as far south as Brittany in France (Maggs & Guiry, 1981). It is difficult to know whether this is a case of migration or whether owing to its small size and position in the sublittoral it had previously been missed. The use of free diving techniques has certainly added much to the knowledge of marine algae around our coasts in recent years.

Codium fragile subsp. *tomentosoides* is certainly a taxon which has migrated fairly rapidly from the south coast northwards since its first discovery at Steer Point in Devon in 1939 (see Silva, 1955). The progress along the south coast of England and the west coasts of Ireland and Scotland can be followed from records in the literature. For the situation in Ireland, see Parkes (1975). The subspecies has now reached the north of Scotland.

A few rare marine species of green algae, e.g. *Cladophora battersia* and

Cladophora retroflexa have recently been found at sites from which they were recorded at the end of the 19th century and they appear to have spread hardly at all. One site for both species is Portland Harbour in Dorset which at present is subject to considerable disturbance and some pollution, and their survival may be threatened. Neither species has been recorded in a reproductive condition. Both are unattached forms growing on small stones or completely free and must spread locally by vegetative means. The other site on the west coast of Ireland is under less threat at present. It is hard to account for this disjoint distribution; it may be a result of the isolation of relict species.

There is strong evidence that pollution of inshore waters is having an effect on the distribution of some green algae. Some species appear to have died out in areas they occupied earlier in the century. *Codium tomentosum* was recorded in the Isle of Man at the turn of the century, but has not been seen there for many years, though it was recorded by Knight & Parke (1931) as frequent. The nearest site for it in the 1960s appeared to be Bardsey Island. The species was found in Dorset in 1959, but has not been found growing there since, though it is plentiful at Beer just over the western border into Devon. The evidence that this species has died out in Liverpool Bay, Dorset, and also in the vicinity of Galway in the west of Ireland, suggests that the operative factor could be pollution. But there is also the fact that in Dorset and Galway, *Codium fragile* subsp. *tomentosoides* has increased considerably, though this is not true of Liverpool Bay. Increase in green marine algae such as *Ulva lactuca* and species of *Enteromorpha* under polluted conditions has been observed for many years.

Using the definition of life history given by Russell (1973) as an organism's 'known morphological, cytological and reproductive phase', no overall pattern of life histories has emerged during the preparation of this account of the Chlorophyceae. A great variety of life histories has been found. However, the full description of the life histories has been hampered by lack of knowledge of the position of meiosis for many species for which the sequence of morphological phases has been worked out. In some cases, authors have made suggestions as to its position which may eventually prove right or wrong. A good example of this is the life history of *Bryopsis hypnoides*, e.g. compare the evidence in Neumann (1969) and Rietema in Neumann (1974). One feature that has emerged is that life histories are on the whole more complex and varied than was suspected in the work on this class in the early part of this century. Fritsch (1935) stated that the majority of the Chlorophyceae were haploid organisms with the zygote the only diploid phase in the life history, meiosis occurring in the germination of the zygote. While this may be true for some, especially freshwater green algae, it is now known not to be true for the majority of marine green algae.

The characteristeric reproductive bodies of the marine green algae are biflagellate gametes, quadriflagellate, occasionally triflagellate, zoospores and multiflagellate zoospores. Biflagellate swarmers may be formed which act as zoospores. Aplanospores and akinetes also occur. One, two or three morphological phases may occur in a life history.

In this volume, life histories are described in terms of sequences of gametophyte and sporophyte phases. The term gametophyte has been used for a morphological (somatic) phase of a life history in which gametes are formed which fuse to form zygotes; the term sporophyte has been used for a morphological phase forming non-fusing reproductive bodies, whether biflagellate, quadriflagellate or triflagellate, or non-motile. The difficulties associated with using these terms have been recognised. Gametophytes may be haploid or diploid, though all too often the nuclear condition is simply not known. Until the association of nuclear phases with morphological stages is fully known, the life history cannot be determined. The occurrence of stages and the sequence in which they occur can be affected by the environmental conditions and may vary from site to site, as, for example in the case of *Capsosiphon fulvescens, Blidingia minima* and species of *Bryopsis*, or even in the same country under different conditions. Species of the genus *Enteromorpha* and possibly of other genera appear to be capable of reproduction in a juvenile condition; also stages in the life history may be by-passed if the plants are subjected to shock, such as sudden low temperature. The life histories of some marine green algae certainly appear to be very flexible, a feature which may be related to the very variable conditions under which they live. Because of the uncertainty of so many of the life histories, no attempt has been made to apply any system of classification to them. Much work remains to be done especially in terms of chromosome counts, in this field.

One type of life history which occurs frequently among the green marine algae is that involving a gametophyte, the gametes of which fuse to form zygotes which develop into unicellular phases, the so-called *Codiolum*-phases because of their similarity, when stalked, to originally described species of *Codiolum*. These form zoospores or aplanospores and frequently appear to represent diploid sporophytes. Such *Codiolum*-phases may, however, be formed from zoospores or even gametes, thus representing accessory phases in a life history. Only when meiosis has actually been seen can such life histories be finally interpreted; the number of cases in which this is known is relatively limited.

As in earlier volumes in this series, the names of orders and families are those used by Parke & Dixon (1976) with references for the families following Silva (1980). The names of genera and species also follow Parke & Dixon (loc. cit.) except for the changes listed below. Synonymy is usually limited to the basionym and other names in the Check-lists of British Marine Algae (Parke & Dixon, 1968, 1976) and Newton (1931). The nomenclature for genera and species in this volume has not been checked completely. To have done this would have delayed publication unreasonably and a compromise had to be reached.

The following name changes should be noted:

Chaetomorpha capillaris (Kütz) Børg. to *Chaetomorpha mediterranea* (Kütz.) Kütz.

Chlorochytrium willei Wright to *Chlorochytrium moorei* Gardner

Endoderma perforans Huber to *Entocladia perforans* (Huber) Levring

Enteromorpha ramulosa (Sm.) Hooker to *Enteromorpha crinita* (Roth) J. Agardh

Epicladia flustrae Reinke to *Entocladia flustrae* (Reinke) Taylor

Heterogonium salinum P. Dangeard to *Pirula salina* (P. Dangeard) Printz

Monostroma fuscum (Post. & Rupr.) Wittrock to *Monostroma obscurum* (Kütz.) J. Agardh

Phaeophila leptochaete (Huber) Nielsen to *Entocladia leptochaete* (Huber) nov. comb.

Phaeophila tenuis (Kylin) Nielsen to *Entocladia tenuis* Kylin

Phaeophila viridis (Reinke) Burrows to *Entocladia viridis* Reinke

Phaeophila wittrockii (Wille) Nielsen to *Entocladia wittrockii* Wille

Rhizoclonium riparium (Roth) Harvey to *Rhizoclonium tortuosum* (Dillwyn) Kütz.

Since the text of this volume was completed, a green marine alga identified as *Microspora filiculinae* P. Dangeard has been found in the Fleet, Dorset by Dr W. F. Farnham. This alga has been recorded previously in the British Isles (Parke & Dixon, 1976) under the name *Tribonema endoozoticum* (Wille) Magne, but as it has now been shown to form starch, it should be included in the Chlorophyceae (see Christensen, 1985; Farnham *et al.*, 1985).

Taxonomic treatment

CHLOROPHYCEAE Wille

CHLOROPHYCEAE Wille in Warming (1884), p. 22.
Chlorophyllophyceae Rabenhorst (1864), p. 2.

Algae of very diverse form growing in a wide range of habitats; majority living in freshwater, but also colonising brackish water and yet others are entirely marine; few in marine plankton, the majority in the benthos; range of morphology very wide including unicells, colonies of cells, coenobia, unbranched and branched cellular filaments, pseudoparenchyma and simple parenchyma, also semi-coenocytes and coenocytes; chloroplast one to many per cell, either a disc, spindle, stellate or plate-like with or without lobes, sometimes perforate or reticulate, either parietal or central in the cells; pigments grass-green with chlorophylls *a* and *b* predominating over carotenes and xanthophylls, proportions approximately as in higher plants; chloroplast enclosed in a double membrane, lamellae forming stacks (thylakoids) of 2–6 or more; starch formed in association with pyrenoids.

Reproduction sexual, asexual and vegetative; reproductive bodies non-motile or motile; motile cells with 2 or 4, occasionally 3, flagella of whiplash type, lacking mastigonemes, occasionally with delicate hairs or scales; flagella usually protruding from a raised papilla at the upper end of the cell.

Life histories very varied.

The Chlorophyceae form an aggregate class of the division Chlorophyta, for which Silva (1980) lists many segregate classes. The actual number of classes in the Chlorophyta is still a matter of discussion at the present time. Three classes, Chlorophyceae, Prasinophyceae and Charophyceae have been recognised in the Check-List of British Marine Algae (Parke & Dixon, 1976) and this has been followed here.

Doubt has been cast on the desirability of maintaining the Prasinophyceae as a class separate from the Chlorophyceae (van den Hoek, 1980) as some of the definitive characters, e.g. scales on the body of the zoospore occur also in the Chlorophyceae though of quite different form. Such scales occur on the zoospore of *Ulothrix zonata* (Sluiman *et al.*, 1980) and of *Monostroma grevillei* (Moestrup, 1978). There are, however, many other features that characterize the class and appear to make it distinct (see Round, 1971).

Two classes, listed by Silva (1980) as segregate classes of the Chlorophyceae, Codiolophyceae (Kornmann, 1973) and Ulvophyceae (Stewart & Mattox, 1978) are founded on a single character or very few characters. The Codiolophyceae is characterized by a heteromorphic life history including a *Codiolum*-phase, either demonstrated to be or presumed to be a sporophyte. The Ulvophyceae is characterized by ultrastructural features of the vegetative and motile cells, especially those demonstrated during cell division. Both of these classes, as pointed out by van den Hoek (in Lobban & Wynne, 1981), cut right across the boundaries based on more obvious morphological characters. Not only that, but the particular characters, from a strictly taxonomic point of view are such that they cannot readily be determined by users of a flora. In any case, the life history of so many green algae is still incompletely known, even if the morphological events are clear, because the cytological details have not been determined, especially the ploidy of the stages. So far, very few taxa, and these mostly freshwater, have been examined for their ultrastructural features. As a result, neither of these classes is ripe for recognition and they are not accepted for the purposes of this volume; such ultrastructural detail may prove to indicate fundamental relationships and may lead eventually to a more realistic and phylogenetic system of classification.

Only six orders are included in the present classification of the marine Chlorophyceae:

> Volvocales
> Chlorococcales (p. 28)
> Prasiolales (p. 18)
> Ulotrichales (including Chaetophorales) (p. 41)
> Cladophorales (p. 135)
> Codiales (p. 179)

The marine Volvocales, being planktonic, do not fall within the scope of this volume which is devoted to the benthic genera.

KEY TO GENERA

1   Thallus microscopic, unicellular 2
    Thallus of macroscopic uniseriate filaments unbranched except in occasional lateral rhizoids; cells uni- or multinucleate 5
    Thallus parenchymatous with cell division in two or three planes, division in the third plane mainly in the formation of branches and during reproduction 8
    Thallus a macroscopic branched uniseriate filament with uni- or multinucleate cells 11
    Thallus of microscopic branched uniseriate filaments, sometimes forming pseudoparenchyma.. 12
    Thallus coenocytic, filamentous; filaments free or interwoven to form a consolidated structure 20
2   Cells endozoic in mollusc shells and calcareous worm tubes, up to 250 × 150 μm, usually with thickened wall extended at one side or end to form one or more simple or complex colourless rhizoidal structures.. *Codiolum polyrhizum** or *Codiolum*-phase of *Monostroma grevillei* or *Eugomontia sacculata*
    Cells epiphytic, endophytic or free-living 3
3   Cells with base prolonged into a colourless stalk, rarely sessile, sometimes in large groups forming a felt on the rock surface, or entangled with filamentous algae .... *Codiolum*-phases**

---

* *Codiolum*-phases of this type occur in the life histories of *Codiolum polyrhizum*, *Monostroma grevillei* and *Eugomontia sacculata* and can, so far, only be distinguished in culture.
** *Codiolum*-phases of this type occur in the life history of species of *Urospora*, *Ulothrix* and other genera and cannot be distinguished without culture work.

Cells stalked, embedded between vertical filaments of the crustose alga
   Petrocelis .... *Codiolum*-phase of *Spongomorpha aeruginosa*
Cells in groups attached to damp walls in caves in the upper littoral
   region .... *Chlorococcum*
Cells embedded in mucilage of colonies of blue-green algae, tubes of
   diatoms or in or between mucilage walls of higher algae .... *Chlorochytrium*
Cells embedded in larger solid red algae such as *Polyides* and *Dilsea*; cells
   with short stalk or papilla ....
   .... *Chlorochytrium*-phase of *Spongomorpha aeruginosa*
Cells solitary, in chains of up to 3 cells or grouped as colonies of 3–4 cells;
   chloroplast parietal with single pyrenoid; reproduction by budding; in
   micro-flora of estuary installations .... *Pirula*
Cells growing on tangled uniseriate or narrow parenchymatous filaments
   of algae 4
4   Cells elongate, stalked, sometimes curved with thick-walled stalk .... *Characium*
   Cells rounded, often with very short stalk .... *Sykidion*
5   Filaments with uninucleate cells; chloroplast plate-like or stellate with or
   without a pyrenoid 6
   Filaments semi-coenocytic; chloroplast a perforated plate or reticulate
   with several to many pyrenoids 7
6   Filaments unattached, short, up to 12 cells long with a tendency to
   dissociate into individual cells; chloroplast lacking a pyrenoid ... *Stichococcus*
   Filaments attached by a basal cell at least when young, soft, gelatinous;
   chloroplast a parietal plate or ring-shaped, with 1–several, rarely more
   than 12 pyrenoids; cells 5–25 (–60) μm diameter .... *Ulothrix*
   Filaments uniseriate when young, unbranched except for single or paired
   rhizoids at intervals; filaments later multiseriate; chloroplast stellate
   with central pyrenoid .... *Rosenvingiella*
7   Filaments prostrate, soft, gelatinous attached by rhizoidal base; cells
   30–300 μm diameter; chloroplast a perforated plate with several to
   many pyrenoids; zoospores with pointed posterior end .... *Urospora*
   Filaments prostrate, entangled, not gelatinous, sometimes with lateral
   single or paired rhizoids; cells cylindrical; mean cell diameter 10–30 μm;
   chloroplast reticulate with many pyrenoids .... *Rhizoclonium*
   Filaments unattached in prostrate or floating masses or attached erect;
   attachment by disc-shaped base of basal cell, supplemented by rhizoids;
   filaments not gelatinous sometimes harsh to touch; cells cylindrical or
   slightly to distinctly barrel-shaped; mean cell width 35–600 μm; chloroplast reticulate with many pyrenoids .... *Chaetomorpha*
8   Thallus of biseriate filaments with cells approximately opposite, prostrate,
   creeping on or in soft mud; chloroplast a narrow parietal ring often
   broadening on the inner wall of cell, with 2–3 pyrenoids .... *Percursaria*
   Thallus multiseriate to solid with cells in packets of 2–many, separated by
   broad mucilaginous walls; chloroplast stellate with one central pyrenoid
   .... *Rosenvingiella*
   Thallus tubular, hollow 9
   Thallus a flat monostromatic or distromatic leafy blade with or without a
   stipe 10

9   Thallus with the wall consisting of vertical rows of cells, often spirally
    twisted; cells in groups of 2 or more enclosed in mucilage sheaths of
    parent cells; branching by separation of vertical rows of cells; chloroplast
    parietal with large single pyrenoid .... *Capsosiphon*
    Thallus arising from a prostrate disc; cells < 10 μm diameter; chloroplast
    stellate with central pyrenoid .... *Blidingia*
    Thallus attached by rhizoids, divisions in the third plane giving rise to
    branches; cells 10 μm or more in diameter; chloroplast a parietal plate
    with 1–several pyrenoids .... *Enteromorpha*
10 Thallus without stipe or with longer or shorter stipe, monostromatic, later
    polystromatic with onset of reproduction; cells in packets of 2 or more
    separated by conspicuous mucilage walls; chloroplast stellate with a
    single pyrenoid .... *Prasiola*
    Thallus monostromatic, sometimes mucilaginous, arising by vertical
    splitting of a primary closed tubular structure; chloroplast a parietal
    plate with 1–several pyrenoids .... *Monostroma*
    Thallus distromatic with shorter or longer stipe; thallus arising by ad-
    herence of sides of primary tubular structure; chloroplast a parietal
    plate with 1–several pyrenoids .... *Ulva*
11 Cells uninucleate or multinucleate; cells of fronds increasing in diameter
    from base to apex with conspicuous apical cells; filament branches of
    three types, erect vegetative, descending rhizoidal and spreading
    curled, sometimes spinous at tip; chloroplast a perforated plate with
    several to many pyrenoids .... *Spongomorpha*
    Cells multinucleate; fronds with cells decreasing in diameter from base to
    apex; all branches similar to main axis except sometimes for occasional
    formation of rhizoidal branches from cells at base of frond; chloroplast
    reticulate with many pyrenoids .... *Cladophora*
12 Thallus endozoic in mollusc shells 13
    Thallus endophytic, embedded in tissues of host plant or endozoic in
    Flustra and other Bryozoa, also in leaves of *Zostera* 14
    Thallus epiphytic forming patches on the surface of the host plant or on a
    muddy substrate or the backs of limpets 17
13 Filaments spreading in the periostracum of mollusc shells .... *Tellamia*
    Filaments radiating beneath the surface or penetrating deeply in calcareous
    shells in submerged habitats or in dead shells washed up on the shore;
    Codiolum–like sporangia formed on filament cells, with or without
    eventual separation from the filament .... *Eugomontia*
14 Thallus with prostrate system of filaments only, creeping in or between
    cells of larger algae, in or between cells of leaves of *Zostera* or in
    Bryozoa such as *Flustra foliacea* 15
    Thallus with both erect and prostrate systems of filaments, creeping
    between cortical cells of larger brown algae, often in *Chorda filum* 16
15 Filament cells with long, often corrugated colourless hairs; cells elongated
    longitudinally .... *Phaeophila*
    Filament cells with or without hairs; hairs when present often with a
    bulbous base; filament cells sometimes irregular in shape .... *Entocladia*
16 Thallus of richly branched filaments sometimes forming a loose pseudo-

parenchyma; erect system of short filaments with end cells extended to form hairs . . . . *Acrochaete*

Thallus mainly of prostrate filaments; erect system represented by bulb-shaped hair-bearing cells seldom separated from prostrate system by further cell division . . . . *Bolbocoleon*

17 Thallus with prostrate filaments adhering laterally to form a pseudo-parenchymatous disc; cells with or without hairs 18

Thallus with erect filaments closely packed to form isolated patches or confluent turfs 19

18 Thallus epiphytic on other algae and on leaves of *Zostera*; each cell of disc with a long colourless hair not cut off from mother cell; chloroplast lobed or irregular with usually 1 pyrenoid . . . . *Ochlochaete*

Thallus epiphytic especially on *Cladophora, Chaetomorpha* and *Rhizoclonium*; long axis of disc cells at right angles to surface of host.*Pringsheimiella*

Thallus epiphytic on other algae and on leaves of *Zostera*; forming rounded discs with long axis of cells parallel with surface of host; disc becoming polystromatic in centre; central cells somewhat rounded, becoming elongated towards periphery . . . . *Ulvella*

19 Forming isolated patches or confluent turfs of closely packed erect filaments with rhizoidal bases penetrating tissues of host (*Fucus* or *Laminaria*) . . . . *Pseudopringsheimia*

Closely packed erect filaments forming a turf, with rhizoidal lower ends; growing in muddy substrate . . . . *Pseudendoclonium*

Erect filaments densely aggregated to form crusts with prostrate filaments spreading over stones and shells, especially limpet shells; tufts mucilaginous; sporangia formed at ends of erect filaments . . . . *Pilinia*

20 Thallus macroscopic, free-living 21

Thallus microscopic, endozoic or endophytic 22

21 Thallus forming soft tufts of loosely branched filaments arising from a prostrate system of interwoven filaments; sporangia as lateral outgrowths of filaments . . . . *Derbesia*

Thallus vesiculate; vesicle up to 10–15 mm diameter . . . . *Halicystis*-stage in life history of *Derbesia*

Thallus with both prostrate and erect filaments, branches of erect filaments distichous or spirally arranged, becoming shorter towards the apex of the axis so that the mature frond is conical in shape or wedge-shaped; base of branch constricted and with internal wall thickening; gametangia formed from ultimate branches . . . . *Bryopsis**

Thallus with interwoven filaments forming a solid structure of a variety of shapes, differentiated into a loose medulla and tightly packed cortex of swollen filament ends (utricles) bearing hairs and gametangia . . . . *Codium**

22 Thallus endozoic forming a branched, sometimes net-like structure in mollusc shells and calcareous worm-tubes . . . . *Ostreobium*

Thallus consisting of swollen multinucleate cells with long hyaline hairs, connected by colourless filaments; cells with many polygonal chloroplasts; endophytic in larger algae . . . . *Blastophysa*

---

* Filaments of both *Codium* and *Bryopsis* may occur as tangled masses; under some conditions the normal *Codium* or *Bryopsis* structures may not be developed. Then it is almost impossible to distinguish them and they can also be confused with *Derbesia* if it is not forming sporangia.

PRASIOLALES West & Fritsch

PRASIOLALES West & Fritsch (1927), p. 164.

Schizogoniales West (1904), p. 98.

Plants occur in both salt and freshwater, usually in sub-aerial situations; young thalli unbranched, uniseriate or multiseriate filaments, with or without rhizoids; mature thalli parenchymatous, solid cylinders or flat, monostromatic blades with or without stipe, attached by rhizoids or unattached; cells arranged in characteristic regular groups separated by broad mucilaginous walls; chloroplast stellate, axile with single central pyrenoid. Reproduction by fragmentation, akinetes and aplanospores; also by an oogamous sexual process with biflagellate male gametes; asexual and sexual plants distinct; vegetative plants diploid, gametes the only haploid phase.

The Prasiolales is a clear cut order with only one family, Prasiolaceae. The only problems presented are those concerned with the separation of taxa within the family.

One family in this order, Prasiolaceae

PRASIOLACEAE (Rabenhorst) Borzi orth. mut., Blackman & Tansley

PRASIOLACEAE Blackman & Tansley (1902), p. 138.

Prasioleae Rabenhorst (1868), p. 307.
Prasiolacee Borzi (1895), p. 237.
Blastosporeae Jessen (1848), p. 13.
Blastosporaceae Wille in Engler (1909), p. 14.

The diagnosis for the family is the same as for the order.

Two genera, *Prasiola* (C. Ag.) Meneghini and *Rosenvingiella* Silva are included in the family. *Hormidium* Kützing, which has uniseriate filaments and *Schizogonium* Kützing, which has multiseriate, monostromatic band-like thalli, are omitted as these are now recognised as developmental stages of *Rosenvingiella* and of some species of *Prasiola*.

Although the family forms a very coherent and easily recognised group of algae, the relationship between the two genera *Rosenvingiella* and *Prasiola*, despite recent experimental work, is still not clear. The idea that *Rosenvingiella* might belong to one or more of the species of *Prasiola* has arisen because frequently it occurs mixed with *Prasiola* species and intermediates seem to occur between them (Knebel, 1935; Waern, 1952; Edwards, 1975). Kristiansen (1972), however, found *R. polyrhiza* to occur in a distinct band above *Prasiola* species in the supralittoral of a shore in Denmark. Edwards (1975) recorded intergrading of *Rosenvingiella polyrhiza* and *Prasiola calophylla* in field samples from Co. Durham and also between *P. calophylla* and *P. stipitata*. Knebel (1935) described the adult flattened *Prasiola* plant as passing through a uniseriate filamentous stage (*Hormidium*) and a band-like multiseriate stage (*Schizogonium*) before reaching the adult form. He distinguished these stages from

Rosenvingiella (*Gayella*) because, in this genus the filaments become solid, the cells of the thalli dividing in three planes. Børgesen (1902) also described uniseriate and multiseriate filaments for *Rosenvingiella polyrhiza* that intergraded with sheet-like *Prasiola*. Because of the resemblance of this *Prasiola* form to *P. crispa*, he re-named *Rosenvingiella* as *Prasiola crispa* subsp. *marina*. This taxon had already been named *P. crispa* f. *submarina* by Wille (1901). Waern (1952) thought that it was incorrect to link the *Rosenvingiella* (*Gayella*) filaments with *Prasiola crispa*, as *Rosenvingiella* has these stages in its own development. Jønsson (1912), instead of following Børgesen (1902) in attaching *Rosenvingiella polyrhiza* to *Prasiola crispa*, brought it into the genus *Prasiola* as *P. polyrhiza*. Newton (1931) gave *Gayella polyrhiza* as a synonym of *P. crispa*. Despite the field evidence of Edwards (1975), Knebel (1935) and Waern (1952), reporting intermediate forms between *Rosenvingiella polyrhiza* and *Prasiola crispa*, *P. calophylla* and *P. stipitata*, there seems no conclusive experimental evidence linking these species. Hooper & South (1977) reported that isolates of *Rosenvingiella polyrhiza* in Newfoundland remained distinct from *Prasiola* under a variety of culture conditions.

Under the circumstances it seems best to keep the genera *Rosenvingiella* and *Prasiola* distinct until such time as clear evidence is available necessitating changes.

There is a similar problem with the species of *Rosenvingiella* and *Prasiola* living on the coasts of North America. *Rosenvingiella constricta* occurs mixed with *Prasiola meridionalis* and, on this ground, Setchell & Gardner (1920) thought that *Rosenvingiella* (*Gayella*) might be a doubtful genus. In culture experiments they were, however, unable to establish any connection between the two. Bravo (1962, 1965) showed that plants resembling *R. constricta* could be formed from the margins of the thallus of *Prasiola meridionalis* and also from the base of the thallus. She claimed that the life history of *Prasiola meridionalis* includes *R. constricta*. Hanic (1979), on the other hand, provided evidence that *Rosenvingiella constricta* and *Prasiola meridionalis* are independent species and he refutes Bravo's evidence. Hanic found the two species to be distinct after examining 5000 plants from 11 generations over a period of 7 years. Kornmann & Sahling (1974) described the life history of *Rosenvingiella constricta* from plants found growing in Helgoland, as a dioecious diplont, meiosis occurring in the zygote giving rise to two female and two male filaments. There was no involvement of a *Prasiola*-type phase in the life history. *Prasiola meridionalis* has not been reported from European coasts, but Bravo (1965) thought that *P. meridionalis* and *P. stipitata* might be conspecific. Although plants of *Rosenvingiella polyrhiza* may sometimes be found with occasional constrictions (Edwards, 1975), plants with the deep constrictions of *R. constricta* have not been found in the British Isles.

PRASIOLA (C. Agardh) Meneghini

PRASIOLA (C. Agardh) Meneghini (1838), p. 360.

Type species: Not designated.

Ulva subgenus *Prasiola* C. Agardh (1822), p. 416.
Hormidium Kützing (1843), p. 244, nom. illeg., *non Hormidium* Lindley ex Heynhold (1841), p. 888.

Frond small, usually more or less rounded, spatulate, or strap-shaped, unattached or attached by rhizoids or by a short multiseriate stalk or longer uniseriate or few-seriate stalk; vegetative frond monostromatic, of cells with long axis perpendicular to plane of frond; cells in groups of 4 or more arranged in regular transverse and longitudinal rows, groups separated by broad mucilage walls; chloroplast stellate, axile with a single central pyrenoid; frond becoming multilayered, parenchymatous during reproduction; asexual reproduction by non-motile spores each formed directly from one vegetative cell, released first into cavity in frond formed by breaking of internal cell walls, later to exterior by disintegration of external walls; lower part of frond remains vegetative; sexual plants with lower part vegetative, upper end becoming multilayered with patchwork pattern of pairs of female and male tissues, each cell forming a single non-motile egg or a biflagellate spermatozoid; eggs and sperms released into thallus cavity formed by breakdown of inner cell walls; fusion takes place in the cavity; release of zygotes by breakdown of fertile part of thallus.

Life history: vegetative thalli diploid, whether spore- or gamete-producing; sexual tissues and gametes haploid; production of spores or gametes depends on environmental conditions. This life history is based on that of *P. stipitata* worked out by Friedman (1959, 1960, 1969). The life histories of the remaining species are not so well known and sexual reproduction has been found in none of them so far. The life histories are either heteromorphic or monomorphic, in either case involving only aplanospore production (Wille, 1906; Knebel, 1935; Kornmann & Sahling, 1974).

KEY TO SPECIES

1   Frond more or less rounded or strap-shaped, lacking stipe, attached by
    rim of frond or by fine rhizoids . . . . *P. crispa*
    Frond slender strap-shaped becoming gradually uniseriate towards base,
    attached by a disc-shaped holdfast . . . . *P. calophylla*
    Frond with short or long stipe 2
2   Stipe ½–1 or more times as long as blade, with 1 or a few cell series; frond
    fan-shaped or lanceolate . . . . *P. stipitata*
    Stipe short, broad and multiseriate; frond fan- or spoon-shaped narrowing
    sharply into stipe . . . . *P. furfuracea*

Prasiola calophylla (Carmichael ex Greville) Kützing (1845), p. 243.

Lectotype: BM–K (See Friedmann, 1959, pl. 3, fig. 8). Lismore Island, Argyll, Scotland.

Bangia calophylla Carmichael ex Greville (1826), pl. 220.

Thallus bright green, narrow ribbon-like, up to 2 cm long, 1 mm broad, tapering to distinct stalk, attached by disc-shaped holdfast; stalk diameter 20 µm; thallus more than 15 µm thick; cells at base in single series, above arranged in distinct longitudinal and transverse rows increasing to 25 or more rows at apex; cells sometimes arranged in small groups; thallus monostromatic in vegetative state, polystromatic on reproduction; cells in stalk 9·0–11·0 × 3.5 μm, quadrate towards apex, 5–6 μm diameter, 8–10 μm high.

Reproduction by aplanospores, 16 per cell formed by anticlinal and double periclinal divisions of thallus cells; spores released by gelatinisation of sporangial walls; spores germinate to form single celled aplanosporangia containing 8–25 spores (Kornmann & Sahling, 1974); reproduction also by akinetes.

Chromosome no.: not known.

Plants found growing at or above high water mark, forming a turf, sometimes mixed with *Prasiola stipitata*; penetrating into estuaries; in Helgoland they occur mixed with *Rosenvingiella polyrhiza*.

Rarely recorded in the British Isles and very few recent records: Durham, Aberdeen, Argyll (Lismore Island), Ulster, Munster, Leinster.

Sweden, Germany, Atlantic coasts of France and Spain.

There is a record for the Pacific coast of North America (Washington), but Scagel (1966) thought that the material concerned belongs to *Prasiola meridionalis*.

Plants can probably be found at most times of the year especially under moist conditions, but there is little precise information concerning either the seasonal occurrence or the reproductive behaviour of this species.

From their culture experiments using Helgoland material Kornmann & Sahling (1974) found the life history to be an obligate alternation of heteromorphic asexual generations, the ribbon-like thallus alternating with a single-celled aplanosporangium.

There is still some doubt as to whether *Prasiola calophylla* is a species in its own right or whether it belongs to *Prasiola stipitata*. This has been commented on by Lagerheim (1883), Waern (1952) and more recently by Edwards (1975). Edwards, in a collection of plants from the tidal limits of the River Wear, Durham, found very elongated thalli of *Prasiola* with a maximum length of 13 mm and a maximum width of 260 μm attached by a uniseriate stipe; the longitudinal walls of the cells were much thicker than the transverse walls. The plants agreed with the description and illustration of Greville (1826) for *Prasiola calophylla*. In the field they were mixed with plants with a wide range of form including some resembling *P. stipitata*. Edwards thought that the elongated thalli and the longitudinal thickenings of the cell walls might be a response to the estuarine habitat. The plants which resembled *P. stipitata* in external form also tended to show an increase in thickness of their longitudinal walls and the numbers of elongated thalli tended to increase with distribution up the estuary. Waern (1952) thought that there was a possibility that *P. calophylla* and *P. stipitata* might be conspecific. Kornmann & Sahling (1974) found no evidence from their culture studies that *P. calophylla* is conspecific with *P. stipitata*.

Prasiola crispa (Lightfoot) Kützing (1843), p. 295.

Holotype: BM(K). The original publication gives no precise indication of the place of collection, although an annotation on the holotype in Lightfoot's hand indicates that it was collected on the Isle of Skye, Argyll, Scotland.

Ulva crispa Lightfoot (1777), p. 972.

Adult thallus bright to dark green, irregular in shape to rounded, sometimes divided or folded, 1–6 (–10) cm long. unattached or attached to substrate by edge of frond or by rhizoidal outgrowths; thalli often grow in dense layers; cells quadrate or rectangular, 6–8 (–13) μm diameter, 13–16 μm high; thalli frequently mixed with uniseriate filaments and all gradations from these to the flat thallus.

Reproduction by akinetes and aplanospores.

Chromosome no.: not known.

The species is found on damp rocks and stones above high water mark and inland on walls, roofs and trees on or below places where birds perch; sometimes within reach of spray from the sea, but usually above this.

Infrequently recorded in the British Isles: Isle of Man, Isle of Skye, Orkney and Shetland Isles, Yorkshire, Co. Galway, Co. Cork.

Spitzbergen south to the Mediterranean, Baltic Sea.

Atlantic coast of North America from the Canadian Arctic south to Quebec, Alaska, Argentina, New Zealand, the Antarctic, Japan.

The species has been recorded especially in the spring, but can probably be found at almost any time of the year if conditions are not too dry. There is no information concerning the period of reproduction.

The life history appears to be an alternation of heteromorphic asexual generations, but whether or not this is obligate is not known. The thalli release akinetes which on germination produce sporangia which dehisce to release aplanospores.

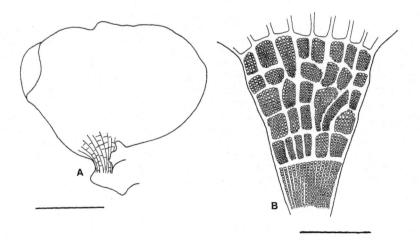

Fig. 2   *Prasiola furfuracea*
A. Habit of plant. Bar= 500 μm. B. Surface view of thallus, basal region, showing quadrate packets of cells arranged in lines. (O. Morton). Bar= 200 μm.

In a population of *Prasiola crispa* there can usually be found narrow almost uniseriate filaments, broader multiseriate fronds and all gradations to the normal adult form. To some of these Knebel (1935) has given subspecific rank.

Prasiola furfuracea (Mertens) Kützing (1843), p. 295.   Fig. 2

Lectotype: original illustration (Hornemann, 1813, pl. 1489) in the absence of material. Hofmansgave, Denmark.

Ulva furfuracea Mertens ex Hornemann (1813), pl. 1489.

Thallus dark green, membranous, up to 2 mm long, fan-shaped spoon-shaped or orbicular, often curled over at edges, as broad as long, grading into short fan-shaped stipe below; stipe sometimes absent, thallus then attached by basal cells or by a few rhizoids; cells often arranged in quadrate packets, in lines fanning out across blade, separated by lines formed of thickened wall of cell groups; vegetative thallus monostromatic.

Reproduction by akinetes, 4–5 μm diameter, formed in thick sporangial layer resulting from parenchymatous divisions of thallus cells; akinetes released by disintegration of thallus.

Chromosome no.: not known.

Plants are found on damp woodwork and in cracks in rock, forming tufts above high water of spring tides.

The species is found mainly in the north of Scotland and all of the recent records are from there: Shetland Isles, Orkney Isles, Aberdeen, Isle of Skye, North Uist, Co. Antrim, 'Connaught'.
Norway, Sweden, Helgoland, Baltic Sea.
Iceland, Faeroes.

The species is known to be present during the summer and early autumn months and it may be present all the year round. Almost nothing is known of its reproductive period. Kornmann & Sahling (1974) found fertile plants in the field in Helgoland at the end of September, a week after a drought of one week's duration. Vegetative plants collected from the field would become fertile after 9–12 days in culture.

There is some disagreement concerning the life history of *Prasiola furfuracea*. It has been reported as a heteromorphic alternation of asexual generations by Wille (1906) and Knebel (1935), but whether obligate or not is uncertain. Kornmann & Sahling (1974) say that the species reproduces only by what they describe as aplanospores, in which the greater part of the thallus is lost. If sporangia can be recognised, they are the mother cells forming aplanospores in the fertile thallus. They did not observe the formation of sporangia on the germination of the aplanospores in their cultures. Knebel (1935) illustrates aplanosporangia developed from spores released from the parent plant. His figures, according to Kornmann & Sahling (loc. cit.) are taken from Wille (1906) and they think that the figures could have been taken from *Prasiola calophylla* and not *P. furfuracea*. Friedmann (1959), however, supported Wille (loc. cit.) and Knebel (loc. cit.).

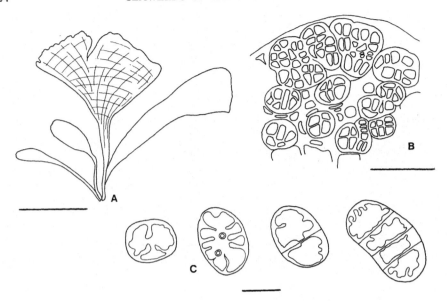

Fig. 3   *Prasiola stipitata*
          A. Habit of plant. (St. Andrews, Fife; H. Blackler) Bar= 500 μm. B. Surface view of
          thallus showing packets of cells. Bar= 50 μm. C. Development of aplanospores.
          (Borth-y-garon; April 1960; Rocks above *Pterocladia* pool). Bar= 10 μm.

Prasiola stipitata Suhr ex Jessen (1848), p. 16.   Fig. 3

Lectotype: original illustration (Jessen, 1848, pl. 2, fig. 11) in the absence of material.
   Sandweick, Schleswig, Germany.

Thallus dark green, wedge-shaped to fan- or heart-shaped, flat membrane up to
1 cm long, narrowing to distinct, sometimes long stipe; cells in rows in stipe,
sometimes a single row at base, in packets of 2, 4, or more in blade; packets of cells
separated by thick walls which form almost straight lines through the thallus; cells
nearly isodiametric at base of thallus, in remainder of thallus the cells are like
elongated prisms with the long axes at right angles to plane of thallus; chloroplast
axile, stellate with processes extending irregularly towards cell wall; pyrenoid single.
    Reproduction by spores and gametes on separate, or exceptionally on the same,
thallus. In spore formation thallus often becomes 2- or 4-layered, one spore per cell
released into cavity between outer lamellae formed by rupture of cells; spore formation proceeds basipetally through thallus. Gametes formed in multilayered haploid
tissue in upper thallus, light tissue forming male gametes, dark tissue forming female
gametes, tissues are formed by divisions following meiosis in thallus cells; gametes,
one per cell, liberated into central cavity formed by breakdown of internal tissues.
    Both spore-forming and gamete-forming thallus diploid; multi-layered haploid
tissue formed following meiosis on sexual plants prior to gamete formation.

Chromosome no.: 2n = 12 (14) Friedman (1959).
     2n = 12–14 Sarma in Godward (1966).
     2n = 16 (*P. meridionalis* Cole & Akintobi, 1963).
     2n = 14 (16) (*P. meridionalis* Bravo, 1965).

Forming tufts on rocks, stones and concrete in the supralittoral fringe and upper littoral zone, penetrating into estuaries; occurs in both exposed and sheltered localities.

Distributed round the whole British coast in suitable habitats.
From North Cape south to Gibraltar.
Iceland, Faroes; Newfoundland and Nova Scotia, Canada; Maine, USA.

Maximum abundance in winter, disappearing in summer in the south of Britain to as far north as the Orkney Islands; north of these islands *Prasiola stipitata* persists through the whole year.

The ratio of sexual to asexual plants, in Anglesey, decreases towards the end of the growing season; no sexual plants are found in most northern localities in either summer or winter. The ratio of sexual to asexual plants is highest at the southern limit of distribution of the species, decreasing northwards. There is also a vertical zonation of sexual and asexual plants, meiotic plants being more prevalent in the lower, and non-meiotic in the higher levels of the *Prasiola stipitata* zone in the upper littoral. There can, however, be exceptions to the general plan of distribution of meiotic and non-meiotic plants (Friedmann, 1959).

Waern (1952) says Sjöstedt (1922, figs 1–24) reproduced (sic) a type series of *P. stipitata* proceeding from broad thalli to narrow cork-screw-shaped thalli which he identified as *P. cornucopiae* J. Ag. Sjöstedt explained the twisted thalli as reduced forms of *Prasiola stipitata* under conditions such as those of: lack of moisture, lowered salinity and variable nutrient content of the water. He maintained the distinction *cornucopiae* as *P. stipitata* f. *cornucopiae* (J. Ag.) Sjöstedt. Waern (1952) remarked that even Jessen's well known figs 11, 7, 15 (plate II) from a Baltic population disclose tendencies to '*cornucopiae*'.

Friedmann (1959) described two morphological forms of spore-producing plants:
1 *Marginal* in which the rate of spore production and liberation keeps pace with the activity of the spore-producing zone so that the spore mass occupies a narrow marginal strip at the apex of the thallus.
2 *Variegated* in which sporogenous activity sets in simultaneously at many points in the thallus: maturing spores tear the anticlinal walls of the lamella to give small pocket-like dilations filled with loose spores and scattered over the thallus.

Transitions between 1 and 2 may occur but are not frequent. The type of spore-producing plant is determined by external factors.

ROSENVINGIELLA Silva

ROSENVINGIELLA Silva (1957), p. 41.

Type species: *R. polyrhiza* (Rosenvinge) Silva (1957), p. 41.

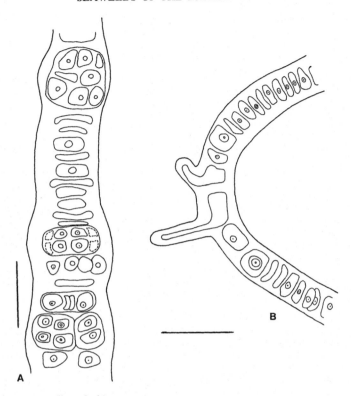

Fig. 4   *Rosenvingiella polyrhiza*
           A. Portion of older, multiseriate thallus. Bar= 20 μm. B. Portion of younger,
           uniseriate filament showing rhizoid development. (Orkney). Bar= 20 μm.

Gayella Rosenvinge (1893), p. 936, nom. illeg., *non Gayella* Pierre (1890), p. 26.

Thallus of uniseriate or multiseriate filaments, attached at intervals by rhizoids, or a cylindrical parenchymatous structure, sometimes with occasional constrictions; cells of larger thalli grouped in packets of 4 or more cells; chloroplast axile, stellate; pyrenoid single.

Reproduction by asexual non-motile spores and by eggs and biflagellate spermatozoids.

The entity was first described by Rosenvinge (1893, p. 936) under the name *Gayella*, but Silva (1957, p. 41) showed that this was an illegitimate name having been used previously in the Anthophyta. He therefore renamed it *Rosenvingiella*. For status of the taxon see discussion (p. 18).

Only one species so far recorded in the British Isles, though a second species, *R. constricta*, has been recorded for Helgoland by Kornmann & Sahling (1974).

Rosenvingiella polyrhiza (Rosenvinge) Silva (1957), p. 41. Fig. 4

Lectotype: C. Godthåb, Greenland.

Gayella polyrhiza Rosenvinge (1893), p. 937.

Thallus filamentous or cylindrical with cells in several superimposed layers, attached by unicellular unbranched rhizoids, often in pairs, sometimes bent; filaments biseriate or multiseriate, (10–) 60–70 µm diameter; axes may be constricted at intervals by undivided cells; rhizoids rare on multiseriate filaments; in multiseriate and cylindrical thalli, cells grouped in packets with cells of two different sizes, 2 or 4 µm diameter; chloroplast stellate, axile; pyrenoid single.

Reproduction by non-motile asexual spores and by gametes; reproductive phase multiseriate or cylindrical, dioecious, larger-celled thalli producing eggs, smaller-celled thalli producing biflagellate spermatozoids.

Chromosome no.: not known.

Found on rocks in the highest littoral region and supralittoral, often mixed with species of *Prasiola*; widespread in estuaries.

The species may be more frequent than is indicated by the records; it appears to be restricted to more northerly coasts. Shetland Isles, Orkney Islands, Fife, Berwickshire, Northumberland, Durham.

Norway, Helgoland, Baltic Sea.

Greenland, Iceland, Faroes, eastern Canada from Labrador to Quebec. The identity of plants from Alaska referred to *Rosenvingiella polyrhiza* is uncertain (see Scagel, 1966).

Plants have been found from January until April, but may be present throughout the year under moist conditions. Edwards (1975a) records the species as perennial in estuaries in Durham. Records of reproduction in the field refer to the spring months. Edwards (1975) observed solid cylindrical plants with cells of two sizes, indicating the onset of reproduction, in February and April, but otherwise nothing is recorded concerning the seasonal reproductive behaviour of this species in the British Isles. Kornmann & Sahling (1974) found fruiting material in Helgoland in the spring.

Kornmann & Sahling (loc. cit.) found as a result of culture studies that the plants are haploid with only the zygote diploid; meiosis was not seen, but equal numbers of male and female plants arose from zygotes, four filaments from each zygote.

The early stages in the development of *Rosenvingiella polyrhiza* are uniseriate filaments, later becoming multiseriate and ultimately in the adult reproductive stage, solid cylindrical structures. Field observations of intermediate forms (Edwards, 1975) have suggested that *Rosenvingiella* is not distinct from species of *Prasiola*, but the culture work of Kornmann & Sahling (1974) has indicated that it is quite distinct (see discussion on p. 19).

CHLOROCOCCALES Pascher orth. mut. et emend.

CHLOROCOCCALES Pascher (1915), p. 2.

Chlorococcoidées Marchand (1895), p. 12, (see Scagel, 1966).
Protococcales (Meneghini) Oltmanns (1904), p. 169.
Protococcoideae Meneghini (1838), p. 4.

Mostly freshwater plants with only a few marine genera; non-motile unicells or non-filamentous coenobial colonies; marine species all unicellular; cells uni- or multinucleate; chloroplast cup-shaped or a parietal plate with lobed or irregular edges, sometimes stellate, entire or reticulate, with one to several pyrenoids.

Reproduction asexual by biflagellate or quadriflagellate zoospores, or by aplano-spores or akinetes; sexual by biflagellate gametes, isogamous or anisogamous, occasionally oogamous.

The Chlorococcales is a large Order of mainly terrestrial and freshwater algae with few marine representatives. A few species occur in brackish water habitats and are extremely rarely recorded in the marine phytoplankton. They range from unicells to large multinucleate coenobial algae. The marine representatives are unicellular and uninucleate in the vegetative state and occur mainly as endophytes of other algae. The systematic position of some of them is uncertain. Some species originally placed here, as for example *Codiolum* and some species of *Chlorochytrium*, have now been recognised through culture work, as phases in the life histories of other green algae, e.g. *Spongomorpha aeruginosa*, *Monostroma grevillei* and species of *Urospora*. Species of the remaining marine genera, *Characium*, *Sykidion* and *Chlorochytrium* may prove to belong to other green algae and perhaps eventually there will be no truly marine representatives of this Order. However, culture work has recently shown that two marine species of *Chlorochytrium* have independent life histories.

Chlorococcaceae is the only family included here.

CHLOROCOCCACEAE Blackman & Tansley

CHLOROCOCCACEAE Blackman & Tansley (1902), p. 95.

Characieae (Nägeli) Wittrock (1872), p. 32.
Chlorochytriaceae Setchell & Gardner (1920a), p. 146.
Endosphaeraceae Klebs (1883), p. 344.
Planosporaceae West (1916), p. 209.
Nautococcaceae Korshikov (1926), p. 491.

Species mainly freshwater, few marine; plants free-living, epiphytic or endophytic, non-motile, often solitary, sometimes occurring together in dense aggregations, unicellular initially, sometimes becoming multicellular; cells spherical, ovoid or angular through pressure, sometimes elongate pedicellate, sometimes attached by a small disc; wall sometimes thick and lamellate or with local thickenings; cell sometimes with a hollow or solid papilla; chloroplast parietal, bell-shaped, spherical, or a lobed plate, sometimes more complex through fusion of broadened ends of lobes; pyrenoids single or few; reproduction by bi- or quadriflagellate zoospores, isogamous gametes, aplanospores or akinetes.

With the removal of the marine species of *Codiolum*, now known to be alternate phases in the life histories of species of *Gomontia, Monostroma, Spongomorpha* and *Urospora*, only four genera with marine representatives remain in the family Chlorococcaceae. Of these, three, *Characium, Chlorococcum* and *Sykidion* have rarely been recorded. *Characium* is largely a freshwater genus with only one marine species. *Chlorococcum* is a terrestrial genus and the one marine species occurs at the upper limit of the littoral region. *Sykidion*, in its morphology, resembles *Chlorochytrium* closely and its status is uncertain. The genus *Chlorochytrium* itself at present includes species which may belong elsewhere and more work is required on these.

CHARACIUM A. Braun *in* Kützing

CHARACIUM A. Braun *in* Kützing (1849), p. 208.

Type species: *C. sieboldii* A. Braun *in* Kützing (1849), p. 208.

Plants epiphytic, unicellular, elongate, ovoid or subcylindrical, attached by a pedicel formed by prolongation of the cell wall; chloroplast bell-shaped or diffuse; pyrenoid single.
Reproduction by zoospores which may be of two sizes, and by akinetes.

Only one marine species in Britain.

Characium marinum Kjellman (1883), p. 317. Fig. 5

Lectotype: original illustration (Kjellman, 1877, pl, 4, fig. 10, as *Characium* sp.) in the absence of material. Musselbay, Spitzbergen.

Characium sp. Kjellman (1877), p. 57.

Cells rounded, elongate, ovate or ellipsoidal, usually with pedicel, c. 40 μm long × 12.5 μm broad; chloroplast bell-shaped or diffuse with one pyrenoid.
Reproduction by biflagellate zoospores which may be of two sizes.

Chromosome no.: not known.

Epiphytic on *Pilayella littoralis* and other algae in the littoral region.

Recorded in Britain at the turn of the century. There is apparently only one recent record, that of Mr Osborne Morton made in 1983 in Co. Down and checked by the author: the species could easily be missed, however.
Berwickshire, Cumbrae, Co. Down.
Spitzbergen, Sweden; Eastern Canada.

Zoospores were found in the Co. Down material in May: they were reported in October in Spitzbergen (Kjellman, 1883). Apart from this nothing appears to be known concerning the seasonal behaviour of the species.
The life history is not known.

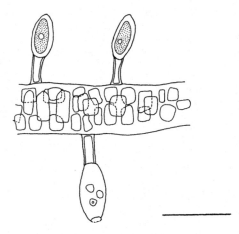

Fig. 5   *Characium marinum*
          Epiphytic habit of two vegetative cells (above) and one reproductive cell (below).
          Bar= 20 μm.

CHLOROCHYTRIUM Cohn

CHLOROCHYTRIUM Cohn (1872), p. 102.

Type species: *Chlorochytrium lemnae* Cohn (1872), p. 102.

Plant unicellular, completely or partially endophytic or occasionally epiphytic in living or dead algae or higher plants, or within the gelatinous tubes produced by some diatoms; cells spherical, ovate or irregular, sometimes curved or lobed; chloroplast a lobed or smooth parietal plate partially or completely filling the cell, lobes may fuse to give fenestrate structure or chloroplast may be solid filling the cell, with fenestrate centre and strands radiating from the central pyrenoid; pyrenoid single or more than one; cell wall variable in thickness, sometimes with internal or external protuberances.

Reproduction by aplanospores, quadriflagellate or biflagellate swarmers.

Despite the removal of *Chlorochytrium inclusum*, now confirmed as part of the life history of *Spongomorpha aeruginosa*, the genus *Chlorochytrium* contains entities the status of which has still to be determined. *Chlorochytrium dermatocolax* and *C. facciolaae* both have features of the '*Codiolum*' morphology and might possibly belong to the life histories of as yet unrecognised algae. There is a great need for the life histories of these two species to be determined.

Apart from the four species included here in the genus *Chlorochytrium*, there are probably others as yet unnamed. Cells having the morphology of *Chlorochytrium* occur as endophytes in a wide variety of algae, but a particular one occurs partially embedded between the superficial cells of *Enteromorpha flexuosa*. This was first identified in *E. flexuosa* in Britain by the present author in material sent by Mr O. Morton (see Morton, 1978). It has since been found in *E. flexuosa* from several other localities and may be the same unnamed *Chlorochytrium* in *Enteromorpha* referred to by Kornmann & Sahling (1977). Of all the *Chlorochytriums* so far described, it appears to show signs of affecting the host plant adversely; the *Enteromorpha* cells surrounding the cells of the endophyte show deterioration of the chloroplast, though this could be a physical effect due to stretching of the host cells. The chloroplast of the endophyte is a lobed cup-shaped plate with smooth edges to the lobes, filling about half of the cell; in this respect it differs from *C. cohnii* which has irregular edges to the lobes of the chloroplast and is quite distinct from *C. moorei* which has a reticulate chloroplast completely filling the cell. More work is required on this *Chlorochytrium* before it is described as a new species of the genus.

KEY TO SPECIES

1 Cells embedded in mucilage tubes of colonial diatoms; chloroplast a
    curved disc or star-shaped with irregular edges . . . .*C. cohnii*
    Only parts of the cells embedded between cells of host   2
2 In fronds of *Blidingia* species; chloroplast fenestrate and filling the cell . . .
    . . . . *C. moorei*
    Between peripheral cells of species of *Sphacelaria* and occasionally of
    species of red algae; cells with a thickened papilla projecting from host,
    sometimes capped by the remnant of the spore wall . . . . *C. dermatocolax*
    Occurring as isolated cells or groups of cells between filaments of blue-
    green algae; cell wall sometimes with projecting thickenings . . . . . *C. facciolaae*

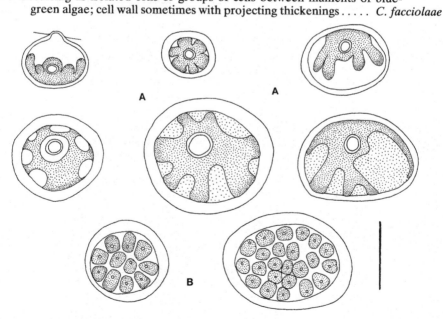

Fig. 6   *Chlorochytrium cohnii*
          A. Habit of vegetative cells. B. Habit of sporangial cells. Bar= 20 μm.

Chlorochytrium cohnii Wright (1877), p. 368. Fig. 6

Lectotype: original illustration (Wright, 1877, pl. V) in the absence of material. Howth, Dublin, Ireland.

Chlorochytrium immersum Massee (1885), p. 301.

Plant unicellular, endophytic; cells rounded, elongate, bottle shaped, or irregular due to compression, with hollow projection up to 4 μm long leading through host wall; cells up to 45 μm diameter, membrane 1.5 μm thick; young cells entering host assuming a figure-of-8 shape; chloroplast a curved disc with smooth or lobed edges or with processes giving an irregular stellate outline, usually occupying only about half of the inner cell wall, sometimes apparently perforate; pyrenoid single.

Reproduction by quadriflagellate zoospores formed by rapid successive division of cytoplasm of vegetative cell which becomes the sporangium; number of zoospores variable, up to about 30 per sporangium; zoospore size variable; biflagellate swarmers also recorded.

Chromosome no.: not known.

Found in tubes of the colonial diatom *Berkeleya rutilans* (Trentepohl) Grün. (also referred to as *Schizonema dillwynii* Ag. and *Berkeleya dillwynii* van Heurck: see Cox, 1975) in marine and brackish water, in rock pools in the littoral region and in the sublittoral; also as a casual in the plankton. The diatom tubes have a wider distribution than the *Chlorochytrium* cells as the cells are more sensitive to drying than the diatom tubes. The species is also reported to occur in some ciliates and coelenterates and a number of brown and red algae, but it has not been proven that the cells in these hosts belong to the same species as those in the diatom tubes (Böhlke, 1978).

Probably more common in the British Isles than is apparent in the records.

Dorset, Devon, Anglesey, Isle of Man, Caernarvon, Isle of Harris, Orkney Isles, Fife, Northumberland, Durham, Sussex, Co. Wexford, Co. Cork.

Sweden, Helgoland, Denmark, France, Mediterranean, Black Sea.

Nova Scotia, Greenland, Iceland.

Probably present throughout the year, but sparse in winter, most plentiful in the summer months. Sporangia have only been recorded in the summer months.

The life history is incompletely known; zoospores germinate directly to form new vegetative cells.

Since Reinhardt (1885) reported biflagellate swarmers released from *Chlorochytrium cohnii* cells and Lagerheim (1884) saw two sizes of swarmers, one biflagellate and the other quadriflagellate, it seemed possible that the life history might also have a sexual phase. Böhlke (1978) suggested that this might be restricted to certain short periods of the year and often missed, while the quadriflagellate swarmers continuously give rise to the asexually formed generation. It is possible that only asexual quadriflagellate swarmers are formed in Britain while in other parts of the world there is a sexual phase, e.g. in Sweden and in the Black Sea where Lagerheim and Reinhardt respectively collected their material. Reinhardt (1885), however, reported that the biflagellate swarmers germinated directly to form new vegetative cells.

Böhlke (1978) gives evidence that Massee's (1885) species, which he found as an endophyte in *Berkeleya rutilans* from Scarborough and named *Chlorochytrium immersum*, is conspecific with *Chlorochytrium cohnii*. Massee observed the release of biflagellate swarmers from the cells and although he did not see fusions between them, he assumed that such had occurred as he saw both smaller and larger cells after the swarmers had settled. If *C. immersum* and *C. cohnii* are accepted as conspecific, it must also be accepted that biflagellate swarmers belong to the life history of the species, though the part they play is uncertain.

Böhlke (1978) found that cells of *Chlorochytrium cohnii* could be grown in culture without the mucilage tubes of the host diatom, but the endophyte grew better in the presence of added carbohydrate and therefore probably benefits by growing in diatom tubes.

Since the above description of *Chlorochytrium cohnii* was completed, Kornmann & Sahling (1983) published an account of the life history of this species based on culture work using Helgoland material. They found that the life history is heteromorphic with the sporophyte a *Codiolum*-like cell with an apparently jointed stalk, growing free of the diatom tubes and producing quadriflagellate zoospores. The latter germinated to produce unicellular gametophyte cells which invaded diatom tubes and when mature produced biflagellate anisogamous gametes: the cells were dioecious. Kormann & Sahling (*loc. cit.*) renamed the species *Chlorocystis cohnii*, the original name of Reinhardt (1885) who described it as forming biflagellate swarmers. They also suggested transferring the species to the Codiolophyceae on the basis of the form of the sporophyte. These results are different from those of Böhlke's culture work in which she found the cells in the diatom tubes to produce quadriflagellate zoospores: these reinfected the diatom tubes without the formation of gametophyte cells. At no time did she see biflagellate swarmers. The life history may be different under different environmental conditions, a state of affairs which occurs in other green taxa, e.g. *Capsosiphon fulvescens*.

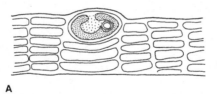

A

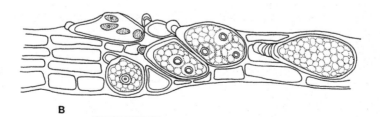

B

Fig. 7   *Chlorochytrium dermatocolax*
         A, B. Endophytic habit of cells in the thallus of *Sphacelaria* (Summer Isles; W.F. Farnham). Bar= 50 μm.

Chlorochytrium dermatocolax Reinke (1889), p. 88.   Fig. 7

Type: KIEL. Kieler Förde, Germany.

Plant unicellular, endophytic; cells somewhat flattened or flask-shaped, 15–50 μm long, 10–40 μm diameter, with stalk-like papilla extending to host surface when the cell is mature; chloroplast in young cells a parietal plate with lobed edges, in older cells lining the whole cell wall, appearing net-like or alveolar, with short processes running towards cell centre; pyrenoid single in young cells, several in mature cells, increase in number probably associated with swarmer formation.

Reproduction by zoospores about 5 μm long, released through a hole in the sporangium wall or through the stalk-like papilla.

Chromosome no.: not known.

Found embedded between cells of peripheral layers of host plant. Reported hosts: *Polysiphonia elongata, P. nigrescens, Sphacelaria racemosa, Rhodomela confervoides, Chaetopteris plumosa.*

Apparently a northern species reported only as far south as Caernarvon in the British Isles.

Caernarvon, Northumberland, Isle of Cumbrae, Ross (Summer Isles).

Sweden, Germany.

Canadian Arctic, Greenland, Iceland.

Probably found through most of the year: swarmers appear to be formed during the late spring and summer months. The type of swarmer formed is not recorded and the life history is not known.

In the early stages of cell formation of the papilla, the zoospore appears to form a germ tube which is not then cut-off from the spore case which later becomes open. Subsequently it becomes thickened and appears as a stalk to the cell much as in *Chlorochytrium inclusum*, now accepted as the sporangial phase of *Spongomorpha aeruginosa*. Svedelius (1901) reported that 'one could clearly observe the hole in the papilla-shaped tip through which the swarmers came out', but he did not make it clear whether he actually saw the swarmers emerge in this way. I have observed empty cells with a hole at the end of the stalk-like papilla. However, in the only cell observed releasing swarmers, they emerged through a hole in the lateral wall of the sporangium. The similarity of the cell to a *Codiolum* phase raises the question as to whether this species is, in fact, a stage in the life history of an as yet unidentified algal species.

I was unable to observe pyrenoids at points in the chloroplast where strands crossed, as described by Svedelius (1901). More than one pyrenoid was observed only in cells already beginning to form swarmers. Bristol (1920) regarded *Chlorochytrium dermatocolax* as a variety of *Chlorochytrium inclusum*. The similarity between the two species was also commented on by Kornmann (1961) though he had not actually found *Chlorochytrium dermatocolax.*

Chlorochytrium facciolaae (Borzi) Bristol (1920), p. 22.   Fig. 8A, B

Lectotype: original illustration (Borzi, 1883, pl. 7) in the absence of material. Messina, Sicily, Italy.

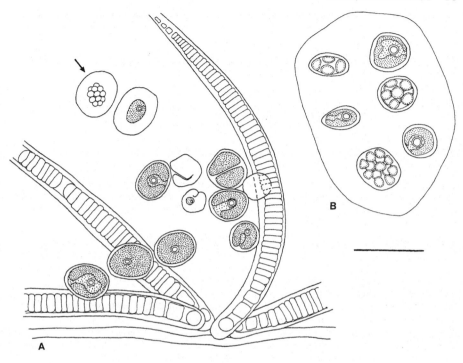

Fig. 8   *Chlorochytrium facciolaae*
A. Endophytic habit of vegetative and reproductive (arrowed) cells within filaments of
blue green algae. (Lough Ine, Co. Cork; T. Norton). Bar= 50 µm. B. *Chlorochytrium
facciolaae* var. *minor* Vegetative and reproductive cells embedded in mucilage. Bar=
20 µm. Both the above taxa were found associated with *Calothrix confervicola*
epiphytic on *Pterosiphonia fruticulosa* (Lough Ine, Co. Cork; 1980; T. A. Norton).

Kentrosphaera facciolaae Borzi (1883), p. 86.

Vegetative cells spherical, ovoid or irregular, 25–45 µm diameter; cell wall thin and
firm, sometimes with small outer projections; chloroplast a parietal plate, sometimes
lobed; pyrenoid single.

Reproduction by aplanospores set free by disintegration of cell wall (Bristol, 1920)
or by zoospores formed in zoosporangia up to 80 µm diameter, with thick wall, 5–10
µm thick, sometimes with internal and external projections; zoospores figured as
biflagellate by Borzi (1883), released through break in sporangium wall or at end of
concrescence.

In the var. *minor* with cells 10–15 µm diameter, the zoospores are held in a sphere of
mucilage.

Chromosome no.: not known.

Cells occur singly or in groups within mucilage of blue green algae or mixed with *Rhizoclonium* on salt marshes, in littoral pools and in the shallow sublittoral; reported also as occurring among blue green algae in freshwater.

The species may be more common than appears in the records, but seems to belong to the southern coasts of the British Isles.

Canvey Island, Cornwall, Co. Cork.

var. *minor* Co. Cork.

Both *Chlorochytrium facciolaae* and *C. facciolaae* var. *minor* were found associated with the same colony of *Calothrix*, var. *minor* in a separate spherical mass of mucilage, in the shallow sublittoral near Lough Ine (Goss-Custard *et al.*, 1979).

Mediterranean, Bavaria.

Massachusetts, USA.; Paraguay.

The seasonal behaviour of this species is not fully known: plants were found in Co. Cork with swarmers in April and July. The life history is not known. Both aplano-spores and biflagellate zoospores are recorded as germinating directly to form new vegetative cells (Bristol, 1920).

Chlorochytrium facciolaae was originally described by Borzi (1883) as *Kentrosphaera facciolaae*. The plants had been found by Dr. Facciola, in association with *Phormidium* and *Oscillatoria*. Borzi included two species in his genus *Kentrosphaera*, *K. facciolaae* and *K. minor*, the latter differing from the former only by the smaller size of its cells and its habitat 'in stagnis submarinis'. The salinity conditions under which the original plants were found are not clear from the description nor were they clear from an early find of *Chlorochytrium facciolaae* in this country, given as 'near Sennen, Cornwall' (Fritsch & West, 1927). A record by Hansgirg (1887) from Bavaria was possibly from freshwater unless saline springs arise in this region. Carter (1933) recorded it from salt marshes in Britain. Recent records in this country show that the species can grow among blue green algae in littoral pools (Goss-Custard *et al.*, 1979). The plants from this habitat, figured here, differ from Borzi's (1883) description in lacking the external thickenings of the cell wall and in having a simpler chloroplast in the form of a thick parietal plate with lobed edges.

Chlorochytrium facciolaae var. *minor* (Borzi) Bristol (= *Kentrosphaera minor* Borzi), recently found near Lough Ine (Goss-Custard *et al.*, 1979), with cells only 10 µm in diameter, has a similar chloroplast. It may be that the morphology of the cells is variable as in *Chlorochytrium moorei* Gardener (= *Chlorochytrium willei* Printz) see Böhlke (1978). There is also the possibility, however, that several different species of *Chlorochytrium* can grow in the habitat offered by the mucilage associated with blue-green algae.

Chlorochytrium moorei Gardner (1917), p. 382.   Fig. 9

Lectotype: BM (Collins, Holden & Setchell, Phyc. Bor.–Amer. No. 565, as *Chlorocystis cohnii* (Wright) Reinh.). Lynn, Massachusetts, USA.

Chlorocystis cohnii sensu Moore (1900), p. 100.
Chlorochytrium willei Printz (1926), p. 219.

Thallus unicellular, endophytic, only partially embedded in host; cells globose in surface view, (22–) 25–35 (–40) µm diameter, funnel or club-shaped in lateral view,

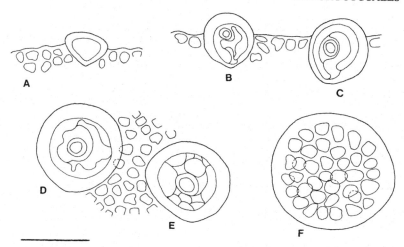

Fig. 9   *Chlorochytrium moorei*
Endophytic habit of vegetative and reproductive cells in *Blidingia*. (Orkney). A-D. cells
of different ages. E. mature cell with alveolate chloroplast. F. Sporangium. Bar= 20 μm.

occasionally with papilla 3–9 μm long; membrane 1.5–4.0 μm thick; chloroplast almost
filling young cells, in older cells fenestrate radiating from a single pyrenoid.
    Reproduction by pyriform quadriflagellate zoospores (5.5–) 7–10 μm long × (4.5–)
5–6 μm broad, containing chloroplast with one pyrenoid and eyespot towards cell
apex; zoospores formed by successive division of mother cell, approximately 150 per
cell, depending on sporangium size.

    Chromosome no.: not known.

    Found partially embedded between the cells of fronds of species of *Blidingia* and
possibly also in other green algae, growing in the upper part of the littoral region; the
density of *Chlorochytrium* cells in the *Blidingia* fronds increases from the upper littoral
down to the midlittoral. The *Chlorochytrium* is absent from *Blidingia* growing at low
salinities and where the organic content of the water is high.

    Generally common round the whole coast of the British Isles from Shetland to the
Channel Isles.
    European Atlantic coast from Scandinavia south to France.
    Newfoundland and the Maritime Provinces of Canada south to Massachusetts,
USA.

    Both vegetative and reproductive cells are present at all times of the year; seasonal
changes in photoperiod and temperature do not appear to influence the stages of the
life history which appears to involve only a succession of asexual generations.

    Chlorochytrium moorei appears to be a distinct species and not part of the life
history of another green alga. In culture Böhlke (1978) has grown it free from the host

Blidingia, but it has not been found free in the field. It seems to be dependent on the host plant for protection against drying. There is some evidence that the *Chlorochytrium* cells are unable to infect *Blidingia* plants with thickened cell walls; for this reason the older parts of the *Blidingia* plants are rarely infected.

There seems no reason now to separate *Chlorochytrium moorei* Gardner from *Chlorochytrium willei* Printz. In 1900 Moore described a unicellular endophyte in *Enteromorpha* on the east coast of America. As Böhlke (1978) has pointed out, although he did not state which species of *Enteromorpha* was the host plant, it would appear from his drawings to have been *Enteromorpha minima* (= *Blidingia minima*). Despite the differences he noticed between his plant and *Chlorocystis cohnii* Reinhardt, in the shape of the chloroplast and the attachment to the host, he assigned it to this species. Gardner (1917) transferred Reinhardt's genus *Chlorocystis* to *Chlorochytrium* Cohn and proposed the name *Chlorochytrium moorei* for Moore's plant, separating it from Reinhardt's plant which he named *Chlorochytrium reinhardtii*. *Chlorochytrium willei* was described by Printz (1926) as a species distinct from *Chlorochytrium moorei* on the basis of its larger cells and the presence of a papilla on them. Böhlke (1978) has shown that both of these characters are extremely variable and unsuitable for use as taxonomic characters. The size of the cells of the endophyte occurring in *Blidingia* species is dependent on environmental conditions and the presence or absence of a papilla is not a constant feature of populations in the British Isles nor on the east coast of America. It has been accepted here that *Chlorochytrium moorei* Gardner and *Chlorochytrium willei* Printz are conspecific.

There is uncertainty as to the identity of the plant referred to *Chlorochytrium moorei*, occurring in *Enteromorpha flexuosa* subsp. *paradoxa*, in the vicinity of the Marine Science Institute, Nahant, Massachusetts (see Webber, 1978).

CHLOROCOCCUM Meneghini

CHLOROCOCCUM Meneghini (1842), p. 24.

Type species: *Chlorococcum infusionum* (Schrank) Meneghini (1842), p. 24.

Cytococcus Nägeli (1849), p. 84.

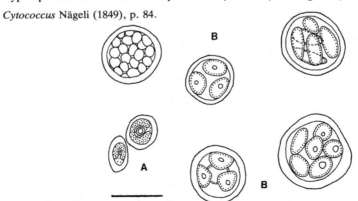

Fig. 10   *Chlorococcum submarinum*
    A. Vegetative cells. B. Reproductive cells. (Drawn from culture; G. Russell). Bar= 10 μm.

Cells uninucleate initially, spherical or angular under pressure; chloroplast parietal, bell-shaped or more or less spherical; pyrenoid single.

Reproduction by biflagellate zoospores, aplanospores, akinetes, gametes and biflagellate.

The name *Chlorococcum* was first used by Fries (1825) but his description was not adequate to identify internal cell structure and Meneghini's (1842) description and illustration of *Chlorococcum infusionum* was the first satisfactory description for the genus indicating the features now accepted.

Only one marine species in Britain.

Chlorococcum submarinum Alvik (1934), p. 28.   Fig. 10

Lectotype: original illustration (Alvik, 1934, pl. 1) in the absence of material. Near Bergen, Norway.

Plant unicellular; cells in culture spherical, 7–10 μm diameter; chloroplast parietal with 1 (–2) pyrenoids.

Reproduction by zoospores produced in sporangial cells 13–16 μm diameter when mature; zoospores biflagellate 6 × 1.5 μm, usually 8 per cell but 4–16.

Chromosome no.: not known.

On the walls of a marine cave, associated with *Pilinia rimosa* and other cave algae.

Hilbre Island, Cheshire.
North Sea.

Little is known about the seasonal occurence and reproductive period of the alga described here, which is the benthic phase of a planktonic alga. It was found in a cave on Hilbre Island, Cheshire, by Rees & Russell (1977) and cultured by them. The reproduction was solely by asexual biflagellate swarmers.

The life history is not completely known.

SYKIDION Wright

SYKIDION Wright (1881), p. 29.

Type Species: *Sykidion dyeri* Wright (1881), p 29.

Plant unicellular, sessile or slightly stipitate; young cells spherical, older cells slightly flattened at apex, sometimes irregularly pentagonal due to mutual pressure; chloroplast cup-shaped, usually appearing to fill cell; pyrenoid single (?). Hamel (1931) records a pyrenoid for *S. droebakense* Wille but no pyrenoid is mentioned by Wright (1881) for *S. dyeri*.

Reproduction by biflagellate zoospores, also by aplanospores.

Of the two species of *Sykidion* described, *S. dyeri* Wright 1881 and *S. droebakense* Wille 1901, only *S. dyeri* is recorded for the British Isles and has not been reported from elsewhere. *S. droebakense* is reported from one or two European countries, but not for Britain. There are only two previous records of *S. dyeri*, both in the last

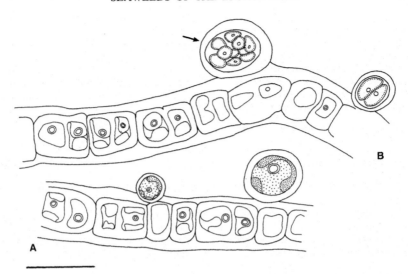

Fig. 11   *Sykidion dyeri*
           A, B. Epiphytic habit of vegetative and sporangial (arrowed) cells on *Ulothrix*
           pseudoflacca. However, note that the illustrated material could belong to the genus
           Chlorochytrium (see text). (Orkney). Bar= 20 µm.

century. The only material found by the present author which approximates to
Wright's description is illustrated in Fig. 11. However, the probability of confusing
Sykidion dyeri with a species of *Chlorochytrium* is very great and the genus needs
reinvestigation.

Sykidion dyeri Wright (1881), p. 29.   Fig. 11

Type: TCD. Howth near Dublin, Ireland.

    Cell bright grass-green attached by narrow base, globose or flattened at apex; cell
wall stiff, persisting after contents released; mature cells 20–25 (–30) µm diameter;
chloroplast usually filling the cell.
    Reproduction by biflagellate zoospores formed in mature cells after formation of an
inner cellulose wall; zoospores released by rupture of outer and inner walls; possibly
also by aplanospores.

Chromosome no.: not known.

    Epiphytic on species of *Rhizoclonium* and possibly also on other mat-forming
filamentous green algae

Howth near Dublin, Ireland; Point of Ayr, Wales; Orkney Islands?.
    This species is not recorded outside the British Isles.

    Present and reproducing in May; seasonal behaviour otherwise unknown. The life
history is unknown.

There is uncertainty concerning the identity of the algal cells identified by Wright (1881) for which he created the new genus and species *Sykidion dyeri*. The only cells seen by the present author which approach Wright's description, attached to filaments of *Ulothrix flacca*, could belong to the genus *Chlorochytrium*. The cells figured by Newton (1931) for *Sykidion dyeri* appear to be 20–30 μm in diameter. The chloroplast was described by Wright (loc. cit.) as 'filling the cell'; Newton (loc. cit.) described it as 'cup-shaped'. The presence or absence of a pyrenoid was not recorded by either author.

Böhlke (1978) has shown that cells of *Chlorochytrium moorei* can develop outside the host *Blidingia* plants on other surfaces. She has suggested that the *Blidingia* plants give protection from drying as would occur on rock surfaces. A mat of *Ulothrix* or *Rhizoclonium* would provide a suitable moist atmosphere and it is possible that cells of this or other species of *Chlorochytrium* might become attached to filaments in such a situation.

A second species of the genus *Sykidion, S. droebakense*, described by Wille (1901) was illustrated as having a lobed chloroplast with a pyrenoid. These cells also have the appearance of a *Chlorochytrium*.

ULOTRICHALES Scagel orth. mut.

ULOTRICHALES Scagel (1966), p. 25.

Ulothrichiales Borzi (1895), p. 348.

Plants micro- or macroscopic, erect, prostrate or heterotrichous, of uniseriate, unbranched or branched filaments, sometimes becoming pseudoparenchymatous or parenchymatous, occasionally with tendency to dissociate into individual cells, attached by basal cell or by branched filamentous prostrate system, or unattached, with filament branches sometimes modified to rhizoidal or specialized hooked branches; parenchymatous forms, unbranched or branched tubes, usually hollow, of two to many vertical rows of cells, or a stalked or sessile, monostromatic or distromatic flat lamina, attached by rhizoids; cells uninucleate, occasionally multinucleate; parent sheaths sometimes clearly persistent through several cell divisions; chloroplast annular or lobed or smooth parietal plate, sometimes perforated, with 1–12 pyrenoids or lacking pyrenoids. Reproduction by biflagellate iso- or anisogametes or by tri- or quadriflagellate zoospores. Vegetative reproduction by fragmentation or by formation of akinetes or aplanospores.

The life history is an isomorphic alternation of gametophyte and sporophyte generations or a heteromorphic alternation with a unicellular stalked or sessile sporophyte; the life history is sometimes complex and the position of meiosis is often not known.

In the Check-list of British Marine Algae of Parke & Dixon (1976), the Order Ulotrichales includes the three families, Ulotrichaceae, Acrosiphoniaceae and Monostromataceae, and is separated from the Ulvales containing the family Ulvaceae and the Chaetophorales containing the families Chaetophoraceae and Chroolepidaceae. As indicated in the discussions under the different genera and species, some authors have split the family Monostromataceae and distributed the genera between the Monostromataceae and the Ulvaceae (Gayral, 1964; Bliding, 1968; Vinogradova,

1969). One difficulty with this is the lack of agreement as to how the separation should be made and, indeed, if all of the characters shown by the plants are taken into account, no distinct taxa at the genus level emerge. It has been decided here to keep the species together within the genus *Monostroma* in the family Monostromataceae and to include both this family and the Ulvaceae within the Ulotrichales. The Chaetophoraceae has also been kept within the Order Ulotrichales rather than in the separate Order Chaetophorales and there seems no basis for placing the marine species of the genera *Pilinia* and *Tellamia* in the family Chroolepidaceae distinct from the Chaetophoraceae. There is fairly convincing evidence that the marine Chaetophoraceae will eventually have to form a new family, the Ulvellaceae, as suggested by O'Kelly (1980), but while the arguments are still in a fluid state and the distinctive morphological characters difficult to define for a work of this kind, it has been decided not to make this change. It is also obvious from ultrastructural studies relating to cell division and the structure of the motile cells (e.g. Pickett-Heaps & Marchant, 1972; Pickett-Heaps, 1975; Stewart & Mattox, 1978; Sluiman *et al.*, 1980; Mattox & Stewart, 1985; Melkonian, 1985; O'Kelly & Floyd, 1985) that the whole classification of the Order based on morphology does not represent phylogenetic relationships and new classifications have been suggested. None of the changes based on ultrastructural characters has been adopted for the purposes of this volume.

The family Capsosiphonaceae has been reunited with the Ulvaceae. It was created by Chapman (1952) to take the genus *Capsosiphon* which contained only two species, *C. fulvescens* and *C. aurea*; the former is widely distributed in the northern hemisphere while the latter is recorded only from New Zealand. The family was separated from the Ulvaceae on two characters: (1) the similarity between *Capsosiphon* and *Prasiola* in the arrangement of the cells in groups of 2 or 4 within the common sheaths of the parent cells, even though the absence of such sheaths was given as a diagnostic character for the species *C. aurea*, a species he described at the same time that he created the family (Chapman, 1952). (The presence of the common sheaths, under some conditions, is not easy to see in *C. fulvescens*; also a grouping of cells in 2 or 4 is a feature which can occur in green algae other than *Capsosiphon* and *Prasiola*, e.g. in *Monostroma oxyspermum*). (2) the supposed absence of motile reproductive bodies. Chapman (loc. cit.) stated that *Capsosiphon* reproduced by aplanospores. The work of Iwanoto (1959), Bliding (1963) and Migita (1967) has shown that certainly *C. fulvescens* reproduces by quadriflagellate or biflagellate zoospores and sometimes also by isogamous biflagellate gametes in some parts of the range of the species.

Bliding (1963) was of the opinion that *Capsosiphon* differs so much from the other genera of the Ulvaceae that its separation into its own family, as suggested by Chapman (loc. cit.) might be justified. If, however, Chapman's description of his *C. aurea* is considered these differences largely disappear and the justification for a separate family Capsosiphonaceae seems unconvincing.

Five families are placed in this order: Ulotrichaceae, Monostromataceae, Acrosiphoniaceae, Ulvaceae, Chaetophoraceae.

ULOTRICHACEAE Kützing orth. mut. Rabenhorst

ULOTRICHACEAE Rabenhorst (1868), p. 298 (as *Ulothricaceae*), 360.

Ulothricheae Kützing (1843), p. 251.

Freshwater and marine algae, attached or free-living, sometimes endozoic; thallus of usually unbranched, occasionally branched uniseriate filaments, attached by specialized basal cell, or lacking such basal cell; filaments sometimes with tendency to dissociate into individual cells; cells uninucleate, occasionally multinucleate, cylindrical, sometimes barrel-shaped; chloroplast cup-shaped or a parietal partial or complete ring, sometimes perforate or reticulate; pyrenoids one to several.

Asexual reproduction by biflagellate or quadriflagellate zoospores; sexual reproduction by biflagellate gametes, isogamous or anisogamous; vegetative reproduction by fragmentation and regeneration.

Four genera are included in the family for the British Isles.

Of the four genera included here in the Ulotrichaceae, only *Ulothrix* and *Stichococcus* belong to the family in the traditional sense of collecting together green algae with unbranched uniseriate filaments with uninucleate cells.

Eugomontia has been transferred from the Chaetophoraceae from which, although it consists of microscopic branched filaments, it differs in having a heteromorphic life history involving a *Codiolum*-like sporophyte and filamentous gametophyte. It also differs in the form and mode of the development of its sporangia from cells of the filaments. *Pseudopringsheimia confluens* also differs from the other members of the Chaetophoraceae in the same way, but has been left in that family for the time being for reasons discussed under the genus *Pseudopringsheimia* (see p. 126). The arguments of O'Kelly (1980) that these two taxa should belong in the Ulotrichales have been accepted, but the present author is unwilling to create a new family in the Order to take *Eugomontia* just now. The whole situation will need reassessment at a later date.

Urospora, usually placed in the Acrosiphoniaceae, has been included in the Ulotrichaceae because of its apparent closeness to *Ulothrix*, although it shares the perforated chloroplast and multinucleate condition of its mature cells of *Spongomorpha*. In the young stage, the uniseriate filaments and the form of the chloroplast make it sometimes difficult to distinguish from *Ulothrix*, and it shares with *Ulothrix* the mode of development of zoosporangia and a heteromorphic life history (see discussion under the genus *Ulothrix* p. 48).

EUGOMONTIA Kornmann

EUGOMONTIA Kornmann (1960), p. 60.

Type species: *Eugomontia sacculata* Kornmann (1960), p. 60.

Thallus endozoic, of irregularly branched filaments radiating under the surface of calcareous shells or penetrating deeply; cells with a single nucleus and with an easily stained substance in the cell walls; chloroplast parietal, with 1–4 pyrenoids.

Reproduction by quadriflagellate zoospores and biflagellate gametes; zoosporangia formed by swelling of cells sunk in substrate; discharge of zoospores through a hyaline tube raised above substrate or hyaline tube absent; gametangia formed on specialized branches projecting out from the shell.

Only one species has been described for the genus.

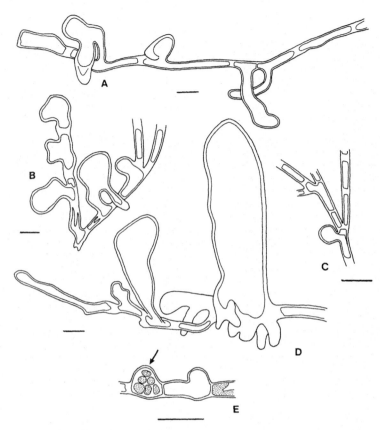

Fig. 12 *Eugomontia sacculata*
A–E. Habit of plants showing enlarged cells and sporangia (arrowed). (M. Wilkinson).
Bar= 10 μm (A, B, D), 30 μm (C), 20 μm (E).

Eugomontia sacculata Kornmann (1960), p. 60.   Fig. 12

Holotype: HELG. List/Sylt, Germany.

Filaments usually 5–10 μm, rarely 2–5 μm diameter; cells 20–110 μm long.
   Zoosporangia spherical, 40–60 μm diameter, or saccate, 60–100 μm × 25–40 μm,
sometimes becoming detached from the filaments; zoospores quadriflagellate,
8–10 μm long; gametangia spherical, 20–40 μm diameter; gametes anisogamous,
smaller 8 μm long, larger about 10 μm.

Chromosome no.: not known.

The plants have been found in a healthy state in a variety of habitats, all permanently submerged, on the beds of shallow, completely flushed estuaries, in permanent pools on tidal sand and mud flats and in the sublittoral. Dead shells containing *Eugomontia* are frequently cast up, but on drying, the plant within the shell becomes shrunken and deformed making identification difficult. Plants have been found only in dead mollusc shells; they are not restricted to any particular species although, by the nature of their habitat, they have been found most commonly in the dead bivalve shells that have been cast up from sandy and muddy bottoms.

Found over the whole coast of the British Isles from the Orkney Islands to the south coast of England; also in Ireland.
Denmark, France.
Atlantic and Pacific coasts of Canada.

Plants are known to be present during the spring and summer months and probably occur all the year round. Although they have been found to produce sporangia and gametangia in culture, very little is recorded concerning their seasonal occurrence in the field.
The life history involves an alternation of isomorphic sporophyte and gametophyte generations, determined in culture (Kornmann, 1960; Wilkinson, 1969).

The morphology of *Eugomontia sacculata* varies with habitat (Wilkinson, 1969). Deep sublittoral specimens growing in shells dredged off Cumbrae in the Firth of Clyde have cells which are relatively long and narrow, as little as 2 μm broad. The cross walls are so indistinct and filaments so narrow that this type can be confused with the siphonous *Ostreobium quekettii* which often occurs mixed with it. Populations from the Isle of Man have been found with normal cell dimensions, but lacking the thickened cross walls. The Manx material differed from the other populations in that the chloroplast was a non-perforated, slightly lobed parietal structure not filling the full width of the cell whereas in other populations it was a parietal perforated plate appearing to fill the cell.
Except in Manx sublittoral material, thickened cross walls have been a useful diagnostic feature being ocasionally as much as 30 μm thick, but more usually about 5 μm. Kornmann (1960) reports that *E. sacculata* can be distinguished readily from other shell-boring filaments because the cross wall stains with zinc chlor-iodide.
Although Kornmann's description states that zoospores are released through special discharge tubes, these have never been found in British material though his description agrees in all other respects.
Although cell dimensions should not be relied on as a definite feature because of their variability, they have been quoted in the description as they help to confirm identification. Plants resembling the zoosporangia, but living free of filaments, were originally described as a separate entity under the name *Codiolum polyrhizum* by Lagerheim (1885). Bornet & Flahault (1888, 1889) claimed that Lagerheim's plant arose by enlargement and separation of cells of a filamentous plant which they called *Gomontia polyrhiza*. However, Kylin (1935) cited evidence from culture that the sporangial *Codiolum polyrhizum*-phase was self-reproducing and independent of filaments. Kornmann (1959, 1960) endeavoured to resolve this problem. From culture experiments he suggested that the following two entities had been confused.
    1 A filamentous plant, boring in shells, which he called *Eugomontia sacculata*, where the filament cells enlarged into a *Codiolum*-like sporangium, but did not separate from the filament.

2 A shell-boring plant resembling Lagerheim's *C. polyrhizum* which was the sporophyte in the life history of a free-living prostrate discoid gametophyte. Kornmann proposed that Bornet & Flahault had both plants in their shells. He did not state clearly the name to be used for the second plant, but appeared to assume that it would be *Codiolum polyrhizum*. In the present work the name *Gomontia polyrhiza* is retained for this plant.

The situation has been further confused by the observation (Wilkinson, unpublished) that, in culture, sporangia of *Eugomontia sacculata* can become detached from filaments and are then morphologically indistinguishable from the shell-boring sporophytes of *Monostroma grevillei* and of *Gomontia polyrhiza*. Also, careful observation of field material of *E. sacculata* from Kent (Wilkinson, unpublished) has suggested that *Codiolum* phases formed from filaments and resembling *Gomontia polyrhiza* have arisen as sporangia on filaments of *Eugomontia*. This means that *E. sacculata* may be very close to the original description of *Gomontia polyrhiza* given by Bornet & Flahault. Certainly it is the case that shell-boring *Codiolum*-phases independent of filaments cannot be assigned firmly to any particular genus or species without recourse to culture work.

STICHOCOCCUS Nägeli

STICHOCOCCUS Nägeli (1849), p. 76.

Type species: *S. bacillaris* Nägeli (1849), p. 77.

Thallus of unbranched uniseriate filaments, lacking special basal cell, with tendency for dissociation into individual cells or groups of cells; filament walls lack sheath; chloroplast a parietal plate usually covering not more than half of the cell wall; pyrenoid, if present indistinct, probably lacking.

Reproduction by fragmentation.

The genus is close to *Ulothrix* in its general morphology though it lacks a distinct basal cell to the filament and also lacks pyrenoids in the cells and any form of motile reproductive body. Nägeli (1849) did not see a pyrenoid in the chloroplast, but one is reported by Collins (1909) and by Taylor (1957), though the latter author says that it is indistinct. More recently Farooqui (1969) has stated that pyrenoids are absent.

Collins (1909) recorded asexual reproduction for the genus by biflagellate zoospores, without stigma, formed singly in the cells and escaping through a hole in the wall. Taylor (1957) reported reproduction by zoospores, by aplanospores and by iso- or anisogametes formed two in a cell or singly. Farooqui (1969) states that *Stichococcus* has no motile cells. Doubts over the presence of a pyrenoid and the formation of zoospores appear to have arisen through confusion between *Stichococcus* and what is now known as *Klebsormidium* (Silva, Mattox & Blackwell, 1972).

Only one marine species, though several in freshwater.

Stichococcus bacillaris Nägeli (1849), p. 77

Lectotype: original illustration (Nägeli, 1849, Tab. IV G) in the absence of material. Zurich, Switzerland.

Cells single or in short uniseriate filaments of 2–8 (–18) cells with rounded ends; cells 2–8 μm diameter, 1.5–5 times as long as broad, cylindrical or ellipsoidal, loosely attached at transverse walls; chloroplast a parietal plate hardly covering cell wall, or lying only on one side and half filling cell or less; pyrenoid indistinct or lacking.
Reproduction by fragmentation.

Chromosome no.: not known.

Found on moist banks and woodwork on the shore and in estuaries and floating in the sea; on vertical rocks down to mid-littoral in places kept wet by water from small fissures in rocks; found also in freshwater.

Rarely recorded in the British Isles: Plymouth Sound, Devon.
North Norway, France.
North American Atlantic coast from Nova Scotia south to New York.

Plants are known to be present in June, but otherwise there is no information on seasonal occurrence or reproduction. The information concerning the species has come from culture work. The life history is uncertain.

This is a freshwater species that can tolerate high salinities and can occur in essentially marine habitats as well as freshwater. Cell size is greater at high salinities than at low salinities.

The species was first described from freshwater by Nägeli (1849) and has usually been treated as a freshwater species since then. George (1957) however, examined three strains of marine *Stichococcus* and found them to agree in shape and dimensions with *S. bacillaris* Näg. The strains were *S. cylindricus* Butcher (Type strain), another strain identified as *S. cylindricus* by Butcher, and Plymouth strain No. 82 (Cambridge No. 379/5) also identified by Butcher as *S. cylindricus* and originally isolated from Plymouth Sound by Parke in June, 1949. George (loc. cit.) found that all of the strains grew well in either fresh- or sea-water media and he therefore concluded that *S. cylindricus* Butcher is synonymous with *S. bacillaris* Näg. He remarked that 'the presence or absence of a pyrenoid cannot be regarded as decisive'.

Haywood (1974), experimenting with Plymouth strain No. 82 in culture, decided that *S. bacillaris* is a freshwater alga that can tolerate high salinities. Also differences in cell size, used as a criterion in distinguishing marine from freshwater species, could be due directly to changes in salinity. There seems no good reason therefore, for separating *S. marinus* (Wille) Hazen from *S. bacillaris* Näg.

ULOTHRIX Kützing

ULOTHRIX Kützing (1833), p. 517.

Type species: *Ulothrix tenuissima* Kützing (1833), p. 517.

Hormotrichum Kützing (1845), p. 204.

Ulothrix forms straight or curved and twisted filaments, unbranched or occasionally branched, attached by a more or less modified basal cell sometimes supported by rhizoids arising from cells above basal cell, or attached by a soft gelatinous ball secreted by wall of undifferentiated basal cell; all cells except basal cell capable of division and reproduction; outer walls smooth or rough; chloroplast ring- or band-shaped with smooth, lobed or toothed edges; pyrenoids one to several.

Asexual reproduction by pear-shaped quadriflagellate zoospores with stigma; zoosporangia cylindrical or barrel-shaped; sexual reproduction by isogamous or anisogamous biflagellate gametes; plants monoecious or dioecious; zygote germinates to form new filaments or forms a unicellular *Codiolum*-phase producing quadriflagellate zoospores; vegetative reproduction by fragmentation or occasionally by akinetes.

Ulothrix is a very widespread genus in both marine and freshwater habitats and is apparently relatively easy to recognise. However, cytological and ultrastructural studies have suggested that the species of *Ulothrix*, which appear very similar morphologically, may not be at all closely related phylogenetically (see e.g. Floyd, Stewart & Mattox, 1972, 1972a; Stewart, Mattox & Floyd, 1973). The situation may have to be reassessed in the future, but for the present, in a taxonomic work of this kind, the species of *Ulothrix* must remain together.

There is sometimes difficulty in separating the species of *Ulothrix* from those of *Urospora* since in young filaments the chloroplast of *Urospora* is very like that of *Ulothrix*. *Ulothrix speciosa* has passed backwards and forwards between the two genera until it was appreciated that some characters, earlier thought diagnostic, such as the pointed base of the motile reproductive body in *Urospora*, can occur in both genera (Lokhorst, 1978). The heteromorphic life history, typical of *Urospora*, also occurs in species of *Ulothrix*.

The number of marine species of *Ulothrix* has recently been reduced by the recognition that *U. consociata* and *U. pseudoflacca* are growth forms of *U. flacca* (Lokhorst, 1978).

Three species of *Ulothrix* are described here for the British Isles.

KEY TO SPECIES

1 Pyrenoid single; chloroplast ocupying only part of cell; cells 5–15 μm
   diameter .... *U. implexa*
   Pyrenoids 1–several 2
2 Filaments forming soft glistening layers, lacquer-like when dry; cells 20–
   50 (–70) μm diameter; outer cell walls with smooth outer layer ....
   .... *U. speciosa*
   Filaments forming woolly masses, dull green when dry; cells 8–30 (–50)
   μm diameter; outer cell walls with rough outer surface often covered
   with particles and small organisms, layer detectable with indian ink or
   iodine in potassium iodide dissolved in lactophenol; filaments sometimes coherent laterally .... *U. flacca*

Ulothrix flacca (Dillwyn) Thuret in Le Jolis (1863), p. 56.   Fig. 13

Lectotype: BM-K (see Lokhorst (1978), p. 213). Swansea, Glamorgan, Wales.

Conferva flacca Dillwyn (1805), pl. 49.

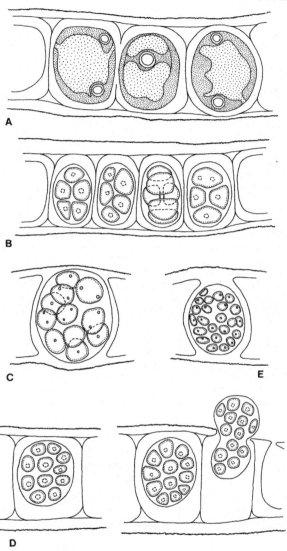

Fig. 13   *Ulothrix flacca*
A. Portion of vegetative filament. B-D. Portions of reproductive filaments, with sporangia. E. Gametangium (note rough outer walls). (Fleet, Dorset; March 1976). Bar= 20 μm.

Gametophyte of flaccid, usually unbranched filaments forming soft woolly masses or flat green layers sticking to substrate, sometimes with a tendency to coalesce, dull green when dry; filaments up to 10 cm long, sometimes tapering to apex; apical cell rounded; filaments straight when vegetative, usually twisted when reproducing; filaments attached by often thick-walled basal cell, up to 100 μm long, sometimes extended below into rhizoids, supported by rhizoids arising from cell above basal cell; filaments uniseriate, occasionally becoming biseriate by longitudinal divisions; filaments with continuous mucilaginous outer layer becoming rough with attached microorganisms and micro-particles. If the wall is smooth, the firm nature of the cell wall can be detected by the use of a solution of iodine/potassium iodide in lactophenol. Cells (4.8–) 14.4–32.6 (–44.2) μm broad, (3.6–) 4.8–9.6 (–15.7) μm long, cell width usually decreasing, cell length increasing towards base; chloroplast band-shaped, parietal, sometimes slightly lobed at margin, usually half ring, occasionally almost complete ring; pyrenoids 1–3 (–8), large globose or ellipsoidal; sporophyte single-celled, more or less globose, stalked '*Codiolum*'-phase 23 μm diameter, or elongated up to 41 μm long × 16 μm broad.

Reproduction of gametophyte by biflagellate gametes, formed in olive brown gametangia, (4–) 8–64 (–128) gametes per sporangium, released through irregular hole in gametangium wall; plants monoecious; gametes isogamous, ovoid to spindle-shaped; also by quadriflagellate zoospores formed in yellow-green sporangia 20–30 × 8–20 μm, from any vegetative cell except basal cell; zoospores ovoid to spindle-shaped, 8–32 per sporangium, liberated in mucilaginous envelope through irregular lateral opening; also by aplanospores 1–32 per cell; vegetative reproduction by fragmentation; reproduction of sporophyte by quadriflagellate zoospores, similar to those of gametophyte.

Chromosome no.: n≈9, 2n≈18 Berger-Perrot (1980b).

Found on rocks and stones and on other algae, also on mollusc shells throughout the littoral region, frequently mixed with other algae, especially *Urospora penicilliformis*, sometimes in pools and on wooden structures on both exposed and sheltered shores; penetrates brackish water of low salinity, found on stems of salt marsh plants, but not usually on flat sand and mud.

Common all round the coast of the British Isles.
Scandinavia south to Portugal, Mediterranean, Black Sea.
American Atlantic coast from South Labrador to New Jersey, Greenland, Iceland, Faroes; American Pacific coast from Alaska south to California; New Zealand; Japan.

Common on British shores in autumn, winter and spring; towards the northern end of the range of the species it is found in spring and summer. In Britain plants will continue on into summer if it does not become too hot and dry.

The reproductive period is mainly in the spring, but will continue into early summer under moist conditions.
The life history is an alternation of filamentous gametophyte and unicellular sporophyte; there is a supplementary cycle of gametophyte generations by quadriflagellate zoospore production; gametes are sometimes parthenogenetic. According to Berger-Perrot (1980b), zygotes may sometimes contain two unfused nuclei in which case they reform gametophytes directly.

Cell diameter is greater in the lower than in the upper littoral region and less in low salinities than in higher.

Lokhorst (1978) regards *Ulothrix pseudoflacca* Wille (1901) and *Ulothrix consociata* Wille (1901) as forms of *Ulothrix flacca*. He has been followed here. Three varieties of the species, var. *flacca*, var. *geniculata* and var. *roscoffensis*, have been described by Berger-Perrot (1980a), on the basis of small differences in morphology and reproduction maintained in laboratory culture.

Ulothrix speciosa (Carmichael ex Harvey in Hooker) Kützing (1849), p. 348.   Fig. 14

Lectotype: BM-K, isotype: TCD. Appin, Argyll, Scotland.

Lyngbya speciosa Carmichael ex Harvey (1833), p. 371.
Urospora speciosa (Carmichael ex Harvey) Leblond ex Hamel (1931b), p. 34.

Thallus of soft bright green glistening unbranched filaments, becoming hard and lacquer-like when dry, up to 10–12 cm long, attached by basal cells; filaments uniseriate, occasionally multiseriate, straight when young becoming curled and twisted when reproducing; cell wall thin with smooth surface, becoming thickened with permanently smooth outer surface, rarely covered by micro-particles or micro-organisms; cell wall sometimes swollen in places or along length of filament; cells cylindrical, isodiametric at filament base; cells above base 0.2–0.25 times as long as broad, (10–) 30–70 (–85) μm diameter; chloroplast parietal, girdle-shaped with 1–3 pyrenoids per cell.

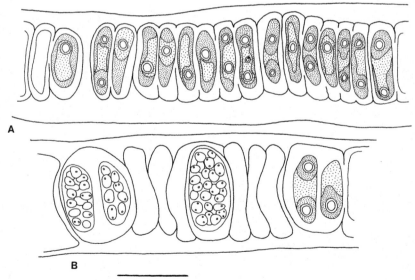

Fig. 14   *Ulothrix speciosa*
A. Portion of vegetative filament. B. Portion of reproductive filament. Bar= 20 μm.

Asexual reproduction by quadriflagellate zoospores; zoospores 7.2–13.8 × 4.2–6.3 μm with pointed base. Sexual reproduction by biflagellate gametes, isogamous or slightly anisogamous formed on filaments separate from those forming zoospores; gametes 5.8–8.6 (–10.4) × 2.9–3.8 μm. The zygote forms a single celled sporophyte giving rise to 32–128 or more quadriflagellate zoospores, or aplanospores; sporophyte up to 75 μm diameter, elongated, sometimes stalked.

The species occurs throughout the littoral region but mainly in the upper part, especially in sheltered areas; often mixed with *Urospora penicilliformis*, *Ulothrix flacca* and *Enteromorpha* species; occurs over a wide variety of substrates including rock, stones and boulders, mud and sand, but only occasionally on fucoids and other algae and animals; found in salt marshes under plants such as *Salicornia*, *Spartina*.

Widely distributed in the British Isles and probably more common than is apparent from the records owing to confusion with other *Ulothrix* species.
Scandinavia south to the Atlantic coast of France.

Plants are abundant in spring, lasting into summer if air conditions remain moist; zoospores are formed over most of this period; gamete production is associated with longer days and tends to decrease at lower salinities (Lokhorst, 1978). The life history is an alternation between a filamentous gametophyte and a unicellular sporophyte.

The cell diameter tends to diminish at lower salinities; the cell length/width ratio is greater in young filaments, but may also persist in older plants growing on salt marshes. A more detailed description of this species and its ecology may be found in Lokhorst (1978).

Of Harvey's (1849*a*) three species of *Lyngbya*, *L. carmichaelii*, *L. speciosa* and *L. flacca*, the first two are based on descriptions by Carmichael. *L. flacca* was placed in the genus *Ulothrix* as *U. flacca* by Le Jolis (1863) and *L. carmichaelii* was considered a synonym, but *L. speciosa* was regarded as a separate species of *Ulothrix*. Hamel (1931*a*) transferred *U. speciosa* to *Urospora* as *Urospora speciosa* (Carm. ex Harv. in Hook.) Leblond ex Hamel (see Edelstein & McLachlan, 1966 and Christensen & Thomsen, 1974), apparently on the grounds that its swarmers were pointed at the posterior end. Kornmann (1964) recognised *Ulothrix speciosa* on the coast of Helgoland and gave an amended description.

Berger-Perrot (1980c, 1981) has added to the confusion by returning *Ulothrix speciosa* to *Urospora* and renaming it *Urospora kornmannii*.

Ulothrix implexa (Kützing) Kützing (1849), p. 349.

Holotype: L 938.174.397 (see Lokhorst, 1978, p. 230). Goes, Netherlands.

Hormidium implexum Kützing (1847), p. 177.
Ulothrix subflaccida Wille (1901), p. 27.
Ulothrix achorhiza Kornmann (1964), p. 37.

Gametophyte of unbranched filaments, solitary, or in tufts or tangled masses, bright green, soft, glossy, up to 5 cm or more long, attached by branched or unbranched rhizoidal basal cell, sometimes supplemented by rhizoids from cells above basal cell; young filaments straight, becoming curled and twisted later; cells usually cylindrical, when mature sometimes with H-shaped remnants of mother cell walls; young cell wall firm and thin, later slightly thickened and rough, occasionally covered by micro-particles or organisms; cells 3.5–26 μm diameter, 3.5–18 μm long, cells becoming narrower towards base of filament; chloroplast in young cells girdle-shaped, extending over $\frac{1}{2}$–$\frac{2}{3}$ cell, sometimes closed by narrow bridge, in mature cells extending towards transverse walls, slightly lobed or toothed along longitudinal margin; pyrenoids 1 (–4); sporophyte unicellular stalked, ovoid or pyriform, up to 70 μm diameter, attached by basal disc; asexual reproduction by quadriflagellate zoospores and aplanospores, formed from any vegetative cell except basal cell; zoospores spindle-shaped or ovoid, 1–8 (–32) per sporangium, released in hyaline envelope through irregular pore in sporangium. Sexual reproduction by biflagellate gametes formed in any cell of a filament except the basal cell; filaments monoecious; gametes isogamous, 2–16 (–32) per cell, ovoid or spindle-shaped, released in hyaline envelope through lateral pore in gametangium; reproduction of sporophyte by quadriflagellate zoospores or aplano-spores, 16 or more per sporophyte; vegetative reproduction by fragmentation of filament and by formation of, sometimes thick-walled, akinetes.

Chromosome no.: not known.

The plants occur mainly in brackish water, on rocks, stones, wooden structures and on other plants, in the upper littoral and a little above this; widespread in estuaries, sometimes mixed with other filamentous algae; they also occur in the mid-littoral where freshwater runs to the sea; also on stems of plants growing on salt marshes.

The species may be more common on the coast of the British Isles than is indicated by the records.

Sussex, Hampshire, Dorset, Devon, Lundy Island, Isle of Man, Cumbrae, Argyllshire, Northumberland, Durham, Yorkshire, Lincolnshire, Co. Cork, Co. Dublin.

Scandinavia south to Portugal, Gulf of Finland.

Greenland, Iceland; California, USA.

The occurrence of the plants appears to be sporadic, but they may be present, even if in small quantity, at any time of the year; most abundant in the spring and early summer. Plants are usually found in a reproductive condition.

The life history, worked out in culture, is an alternation between a filamentous gametophyte and a unicellular sporophyte (Lokhorst, 1978). The life history of plants in Japan has been confirmed as similar to that of plants in Europe (Ohgai et al., 1975; Ohgai & Fujiyama, 1976).

Aplanospores are formed particularly under long day conditions. In nature a prolonged inundation by freshwater decreases the ability of the plants to produce zoospores. Growth of the alga in a medium with a salinity of 2–16‰ and a long photoperiod was found to be necessary for sexual reproduction. Maturation of sporophytes was only achieved after transferring non-fused gametes or zygotes into short day conditions (Lokhorst, 1978).

Setchell & Gardner (1920) followed Hazen (1902) in placing *Ulothrix subflaccida* Wille as a synonym of *U. implexa* Kützing. Hamel (1930) included *U. implexa* in the synonymy of *U. subflaccida* along with *U. cutleriae* Thuret in Le Jolis (1863) and *Lyngbya cutleriae* Harvey (1851). Wille (1901) did not think that his *U. subflaccida* could be identified with *U. implexa* Kützing as it was totally marine whereas *U. implexa* belongs to brackish or freshwater. Perrot (1972) described, under the name *U. subflaccida* Wille, a plant from an estuarine habitat which she thought corresponded with *Lyngbya cutleriae* Harvey (= *Ulothrix cutleriae* Crouan & Thuret in Le Jolis). From this it would appear that the distinction between *U. subflaccida* Wille and the remaining species mentioned above is based mainly on habitat, *U. subflaccida* being truly marine and the others belonging to more brackish habitats. It is not easy to distinguish them morphologically. Lokhorst & Vroman (1974), in an account of the morphology, reproduction and ecology of *U. implexa* (Kützing) Kützing, stated that the species was insensitive to salinity, has a broad ecological range and that the morphological features were hardly changed in the high and low salinities of their culture media. On this evidence there would seem to be no grounds for separating *U. subflaccida* Wille and *U. implexa* Kützing. In a later paper (Lokhorst, 1978), *U. subflaccida* and *U. implexa* are separated as distinct species. The differences between them appear to be few, though presumably constant, and there is a degree of overlap between them particularly in cell dimensions. There is at present little information as to how the few characters separating the two species are inherited, so that it is difficult to assess their value as specific characters. So far, in this country, they have not been separated and, in view of the difficulties in distinguishing them morphologically, it is more convenient to keep them together for the time being under the older name of *Ulothrix implexa* (Kützing) Kützing.

UROSPORA Areschoug nom. cons.

UROSPORA Areschoug (1866), p. 15.

Type species: *Urospora mirabilis* Areschoug (1866), pp. 16, 20.

Hormiscia Fries (1835), p. 327.

Four species of *Urospora* have been described for western Europe by Lokhorst & Trask (1981), *U. wormskioldii* (Mertens in Hornemann) Rosenvinge, *U. bangioides* (Harvey) Holmes & Batters, *U. penicilliformis* (Roth) Areschoug and *U. neglecta* Kornmann. It is uncertain how many of these occur in the British Isles. *U. wormskioldii*, with cells large enough to be seen with the naked eye, would not easily have been missed, but has apparently not been recorded since a single record in 1894. *Codiolum gregarium* and *C. pusillum* have been reported a number of times, but whether or not these belong to the life history of *U. wormskioldii* is uncertain. The characters used to identify the three remaining species of *Urospora* overlap and are not easy to use for field material. Lokhorst & Trask (loc. cit.) maintain that culture experiments are necessary to be certain of the identifications. *U. neglecta* Kornmann has so far not been reported for this country and, although there are several records of *U. bangioides*, the present author has been unable to separate the latter from *U. penicilliformis*, though no critical culture work has been carried out. Kornmann (1966) regarded *U. bangioides* as a growth form of *U. penicilliformis*. There remains the question as to whether the differences found in culture by Lokhorst & Trask (loc. cit.), though apparently constant, are sufficient to warrant the separation of the three species. Their ultrastructural studies did not support such a separation.

Leaving open the question of the identity of *U. bangioides*, only two species of *Urospora*, *U. penicilliformis* and *U. wormskioldii*, have been included here.

KEY TO SPECIES

Cells 25–80 (–100) μm broad; cells cylindrical, only slightly constricted at cross
    walls when mature .... *U. penicilliformis*
Cells (80–) 100–500 (–1200) μm broad; mature cells barrel-shaped .. *U. wormskioldii*

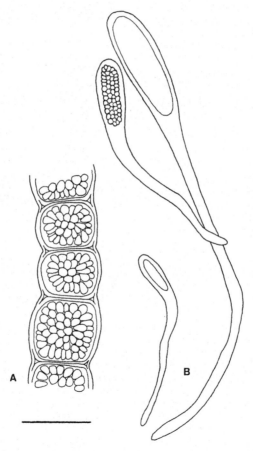

Fig. 15 *Urospora penicilliformis*
    A. Portion of reproductive filament, with sporangia. B. *Codiolum* phase. (Osmington, Dorset; July 1974). Bar= 50 μm.

Urospora penicilliformis (Roth) Areschoug (1874), p. 4, pl. 1, figs 1–6.　　　Fig. 15

Isotype: LD. Nordernly, Germany.

Conferva penicilliformis Roth (1806), p. 71.
Urospora isogona Batters (1902), p. 14.
Hormiscia penicilliformis (Roth) Areschoug (1866), p. 12.

Plant with filamentous gametophyte and unicellular sporophyte; filaments of gametophyte unbranched, 2–8 cm long, 25–80 (–100) μm broad, attached by basal cell reinforced by rhizoids arising from cells above basal cell; cells isodiametric, slightly constricted at cross walls in reproductive condition; chloroplast a perforated plate, sometimes spirally wound; pyrenoids several per cell; sporophyte with cell base prolonged into stalk 300–400 μm long; stalk becoming spirally twisted and rucked or wrinkled; chloroplast with several pyrenoids.

Reproduction by quadriflagellate zoospores formed in cells of filaments and in sporophyte, zoospores radially arranged in sporangium; also by biflagellate gametes from filament cells, anisogamous, male gametes 6–7 μm long, female average 12 μm long; plants dioecious; male gametes released in a membrane.

Chromosome no.: n = 4 Jorde (1933); however, Hanic (1965) suggested that her material belonged to *U. wormskioldii*.

The plant forms soft tufts on rocks and timber at the uppermost littoral level, also sometimes in the midlittoral and possibly in the lower littoral; they are often mixed with *Ulothrix flacca*, *Ulothrix speciosa*, *Blidingia minima* and *Bangia atropurpurea*.

The species is distributed round the whole coast of the British Isles.

European Atlantic coast from Scandinavia south to Portugal, Baltic Sea, Black Sea. Japan, Ochotsk Sea, Bering Sea, Arctic Ocean, American Pacific coast from Alaska to central California; American Atlantic coast from the Canadian Arctic south to New Jersey; Greenland; Iceland, Faroes.

The species is found in winter, spring and sometimes on into summer, depending on the season, with maximum development and reproduction in early spring in the British Isles; the sporophyte has been found along with sexual filaments in February in what appeared to be a pure stand of the species; abundant in spring and summer in northern Canada, early spring in Japan, December to March in Helgoland, collected only in March in Portugal.

The life history is an alternation between a filamentous gametophyte and a unicellular sporophyte; both male and female gametes can develop parthenogenetically forming unicellular stages.

Jorde (1933) recorded the presence of dwarf filaments and also *Codiolum gregarium* as obligate stages in the life history of *Urospora mirabilis* (= *U. penicilliformis*). The dwarf filaments were not seen by Kornmann (1961a) who grew *U. penicilliformis* through its life history. Also he found that *Codiolum gregarium* belongs to the life history of *U. wormskioldii*. Kornmann (1961a) recorded *Codiolum*-like sporophytes in the life history of *U. penicilliformis* and remarked that these had not been found in nature, though Lokhorst & Trask (1981) later refer to naturally occurring sporophytes. The present author found an apparently pure stand of *U. penicilliformis* on a pier at Weymouth with fertile filaments and *Codiolum* stages mixed together. The *Codiolum* was much smaller than that described for *Codiolum gregarium* A. Braun.

Urospora wormskioldii (Mertens ex Hornemann) Rosenvinge (1892), p. 57.

Type specimen: original illustration (Hornemann, 1816, pl. 1547), in the absence of material. Godthåb, Greenland.

Conferva wormskioldii Mertens ex Hornemann (1816), p. 8.
Urospora collabens (C. Agardh) Holmes & Batters (1890), p. 73.
Codiolum gregarium A. Braun (1855), p. 19.
Codiolum cylindraceum Foslie (1887), p. 189.

Thallus of three different kinds, filamentous, dwarf filamentous and unicellular *Codiolum*-like; filaments flaccid, gelatinous, glossy in mass, up to 10–16 cm long, (80–)100–500 (–1200) μm broad, tapering towards base, attached by basal cell reinforced by rhizoids arising from cells above the basal cell, descending closely applied to lower cells or intramatrical; cells multinucleate, 1–1.5 (–3) times long as broad, somewhat barrel-shaped above, cylindrical quadrate towards base; chloroplast parietal, cylindrical in large cells, coarsely reticulate with many pyrenoids; dwarf plants with random cell divisions forming clusters of rounded cells attached to substrate by cell bases and/or rhizoids; cells with central vacuole; chloroplast parietal, finely reticulate with many pyrenoids; unicellular *Codiolum*-like plants elongate, up to 70 μm broad, 270 μm long with base prolonged into a hyaline stalk; chloroplast parietal, seldom reticulate, with many pyrenoids.
    Reproduction by biflagellate gametes, formed in filament cells, anisogamous, plants dioecious and by quadriflagellate zoospores formed in all three types of thallus.

Chromosome no.: n = 12 (Hanic, 1965).
    n = 4 (Jorde, 1933 if her material was *U. wormskioldii* as has been suggested by Lokhorst & Trask, 1981).

The filamentous phase is said to occur on rocks in the littoral region and on submerged rock and wood surfaces. The *Codiolum*-phase referred to *Codiolum gregarium* and *Codiolum pusillum*, forms, in quantity, soft dark green turfs in the uppermost region where it may be wetted at every tide or left exposed for several days at a time.

There does not appear to have been any record of the filamentous phase of this species in the British Isles since 1894, though the *Codiolum* phase has been found a number of times.
Dorset, Devon, Isle of May, Northumberland, Orkney Islands.
Spitzbergen, Scandinavia south to France.
Pacific coast of North America, from Alaska south to California; American Atlantic coast from the Canadian Arctic south to New Brunswick and Maine; Greenland; Iceland, Faroes.

There is insufficient information to make an assessment of the seasonal behaviour of this species. In British Columbia, Hanic (1965) found sexual plants from early April to late September. Very large asexual plants were found at low tide in the same place in spring and early summer (April to June) in two successive years at one locality. Hanic records a short life 10–20 days for the plants. He found *Codiolum gregarium* all the year round, the plants becoming fertile in the autumn and reaching a peak of fertility in December and January. In Norway, Sundene (1953) recorded *Urospora wormskioldii* in the spring and *Codiolum gregarium* in winter and early spring.

The life history is complex and affected by environmental conditions, but is fundamentally a heteromorphic alternation of filamentous gametophyte and *Codiolum gregarium* sporophyte, each phase able to reproduce itself and to give rise to dwarf plants producing quadriflagellate zoospores (Hanic, 1965). The actual position of meiosis in the life history has yet to be determined.

As has been discussed under *Urospora penicilliformis* (see p. 56), there is still doubt concerning the life history to which *Codiolum gregarium* A. Braun actually belongs.

Hanic (1965) found that environmental conditions had an effect on the formation of filamentous and dwarf plants in the life history. The numbers of dwarf plants increased both at higher and lower salinities and were predominant at 10‰ and 50‰; filaments predominated at 20–30‰. At low light levels the number of dwarf plants increased while at higher light intensities normal filamentous plants were formed. Dwarf plants formed more rhizoids at low salinities (10‰) and none at high salinities (50‰). At these high salinities they frequently produced large, up to 200 μm diameter, thick-walled cells.

MONOSTROMATACEAE Kunieda

MONOSTROMATACEAE Kunieda (1934), p. 106.

Adult thallus monostromatic, flat lamina arising from a closed tube or sack by splitting, attached by basal cells or rhizoids; lamina entire or split into shallow or deep segments, margins flat or wavy; cells arranged regularly or irregularly; cell walls sometimes gelatinous; chloroplast a parietal plate with one or more pyrenoids.

Reproduction by biflagellate gametes and quadriflagellate zoospores. Life history an isomorphic, or heteromorphic alternation of generations with a unicellular *Codiolum*-phase sporophyte.

The family Monostromataceae collects together mostly marine and brackish water green algae with a leaf-like monostromatic thallus and is almost certainly an artificial assemblage of taxa. It includes the genera *Gomontia* and *Monostroma*, the affinities of both are uncertain. Kunieda (1934) separated the Monostromataceae from the Ulvaceae on the assumption that its members have a heteromorphic life history while those of the Ulvaceae have an isomorphic alternation of gametophyte and sporophyte generations. With the acceptance here of *Monostroma* as a genus including all the species originally described under this name and also under the name *Ulvaria*, the genus includes species with several different types of life history. *Gomontia* fits uncertainly along with *Monostroma* in the family, though having some similarities to it. *Gomontia* has a shell-boring '*Codiolum*' sporophyte stage in its life history which occurs with and is difficult to separate morphologically from the shell-boring '*Codiolum*' sporophyte stage of *Monostroma grevillei*. This is not, however, as has been suggested by Chihara (1962), an assured guide to their affinity. '*Codiolum*' sporophyte stages are now known to occur in genera belonging to several families in the Ulotrichales. The species of *Monostroma* (see p. 62 under the genus), apart from having the monostromatic leaf-like morphology, may have few characters in common. A number of attempts have been made to separate them into three or four genera, some of which would appear to fit better into other families. Until such time as agreement is reached on how they should be classified it would seem better to keep them together in the family Monostromataceae, even though several species do not satisfy Kunieda's (1934) original description of the family.

GOMONTIA Bornet & Flahault

GOMONTIA Bornet & Flahault (1888), p. 164.

Type species: *Gomontia polyrhiza* (Lagerheim) Bornet & Flahault (1888), p. 163.

Plants with two distinct phases: sporophytic phase unicellular, boring in calcareous substrata with cell wall thickened and extended on one side into 'rhizoidal' structure with layered wall material; gametophytic phase a microscopic multicellular disc, not shell-boring, parenchymatous or pseudoparenchymatous, becoming multilayered at centre.

Reproduction by quadriflagellate zoospores released from sporophyte either through special hyaline discharge tube or by rupture of sporangium; also by biflagellate gametes formed in gametangia at centre of gametophytic disc, forming diffusely on larger plants, or in a ring at centre of smaller plants.

The existence of only one species, *Gomontia polyrhiza*, has been established for Kornmann's (1959) emended concept of the genus, although a second species, *G. manxiana*, was described for Britain (Chodat, 1897) based on Bornet & Flahault's concept of the genus. Wilkinson (1975) has suggested that *G. manxiana* Chodat may be conspecific with an *Entocladia* species he found in shells (Wilkinson, 1969), but this requires further investigation.

Gomóntia polyrhiza (Lagerheim) Bornet & Flahault (1888), p. 163.   Fig. 16

Type: S. Kristineberg, Sweden.

Codiolum polyrhizum Lagerheim (1885), p. 21.

Sporophyte unicellular, up to 250 × 150 μm or much smaller; rhizoid simple, relatively long and pointed or a stout blunt projection with conspicuous layering from cells longer than broad, or branched with a few blunt branches or many finely branched extremities without conspicuous layering arising from cells usually shorter than broad; chloroplast appearing to fill most of cell, parietal reticulate or perforated plate; gametophyte multicellular, parenchymatous or pseudoparenchymatous; parenchymatous forms about 100 μm diameter, of cells 4–10 μm diameter, multi-layered at centre tapering to one layer at edge, sometimes rounding up in older plants to more globular form and becoming multilayered at edge, whole gametophyte often covered by a mucilaginous envelope; chloroplast parietal with one pyrenoid. Gametophyte development sometimes commences as a short filament giving rise pseudodichotomously to a disc multilayered at the centre, but with free filaments projecting from the single-layered edge. Development sometimes continues in an irregular fashion to give an amorphous mass of cells several mm in diameter, which can become free-floating by detachment from the substrate, gametangia forming at any point in the thallus rather than in a ring in such large forms.

Zoospores quadriflagellate, 7–9 × 4–5 μm, released either by rupture of cell or

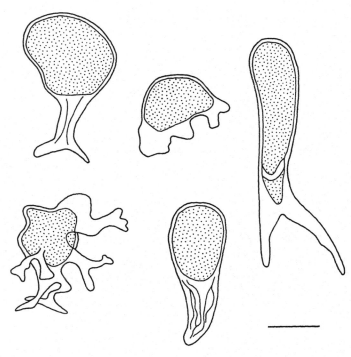

Fig. 16   *Gomontia polyrhiza*
           Unicellular sporophytes. (Figures drawn by M. Wilkinson). Bar= 30 μm.

through special discharge tube. Gametangia forming progressively from centre of gametophyte outwards so that plant becomes ring-shaped; monoecious; gametes 6–9 μm long.

Chromosome no.: not known.

The sporophytes (of uncertain specific origin) have been found in a wide range of species of mollusc shell and in calcareous worm tubes, both living and dead, from the sublittoral through to the littoral, in rock pools and on open rock surfaces, on salt marshes and sand flats and in the lower reaches of estuaries. Gametophytes are known mainly from cultures.

Since the sporophyte is morphologically indistinguishable from similar phases in the life history of *Monostroma grevillei* and *Eugomontia sacculata* and the gametophyte in Britain is known only from culture, the confirmed distribution of this species can only include those localities from which sporophytes gave rise to gametophytes in culture: Anglesey, Cheshire. However, shell-boring *Codiolum* phases resembling the sporophyte, seem to be of universal occurrence around the whole coast of the British

Isles, often where there is no known population of *Monostroma grevillei* or *Eugomontia sacculata* close at hand. It is therefore likely that *Gomontia polyrhiza* is very common in Britain.

Scandinavia south to Portugal, Black Sea; Helgoland; Faroes, Iceland; Greenland, Atlantic coast of Canada and U.S.A.; Pacific coast of Canada and U.S.A.; Caribbean, Canary Isles; Japan, Australia, New Zealand.

Sporophytes have been reported in a fertile condition in March and through the summer months to October and November, but the full seasonal behaviour of the species is not clear. The life history, as observed in culture, is an alternation between a single-celled sporophyte and a filamentous or disc-shaped gametophyte. Wilkinson (1969) presented evidence that the alternation may not necessarily be obligate.

The gametophyte was initially discovered in culture by Kornmann (1959) from German material and has been confirmed for British material in cultures established from sporophytes collected in the field by Wilkinson & Burrows (1972). The only published field record of the gametophyte is that of Pedersen (1973) from Greenland, but Kornmann (pers. comm. to Dr Wilkinson) has also since found it in the field in Helgoland. The life history has been confirmed as similar in German and British material.

The description of the species presented above includes, under the name *Gomontia polyrhiza*, all those morphological types of sporophyte that Setchell & Gardner (1920*a*) proposed as differentiating four separate species of *Gomontia*. Kornmann (1959) has shown in culture that the morphology of the sporophyte varies depending on the culture conditions and Wilkinson (1969) has shown that all of the types and intermediates of Setchell & Gardner (1920*a*), can be produced from one set of parents in culture. Wilkinson has also shown that the sporophyte of *Monostroma grevillei* can produce forms resembling more than one of the species of Setchell & Gardner and that a determining factor in the morphology of the *Monostroma* sporophyte is the particular host species.

Wilkinson (1969) has also reported various forms and types of development for the *Gomontia* gametophyte but, similarly, it would be unwise to use this as a character for specific distinction, especially in view of the polymorphism of prostrate discoid forms in the Chaetophoraceae demonstrated by Yarish (1976).

Kornmann (1962) noted the similarity between the early development of the gametophyte of *Monostroma grevillei* and the gametophyte of *Gomontia polyrhiza*. It is possible therefore that *G. polyrhiza* could represent a form of *M. grevillei* in which full development of the gametophyte to the erect macroscopic phase is arrested by environmental conditions and which becomes prematurely reproductive. Clearly much work remains to be done on the possibility that a number of free-living green algae may have shell-boring stages in their life histories and that the shell-boring stages may be much more difficult to distinguish than the free-living forms.

MONOSTROMA Thuret

Monostroma Thuret (1854), p. 29.

Lectotype species: *Monostroma oxycoccum* (Kützing) Thuret (1854), p. 29.
   [= *M. oxysperma* (Kützing) Doty (1947), p. 12].

Adult gametophytic thallus a monostromatic, flat lamina, arising from a closed tube or sack, attached by basal cells supplemented by rhizoids arising from cells above basal disc; lamina entire or split into shallow or deep segments, margin flat or wavy; cell walls thin, sometimes gelatinous; cells arranged regularly or irregularly, uninucleate; chloroplast a parietal plate; pyrenoids one or more; sporophyte a flat monostromatic lamina similar in form to a gametophyte or a unicellular stalked *Codiolum*-phase.

Reproduction by biflagellate gametes and quadriflagellate zoospores.

The name *Monostroma* Thuret (1854) is an illegitimate name for an assemblage of green algae all of which have a leaf-like thallus consisting of a flat monostromatic layer of cells, formed either by upwelling of a vesicle from a prostrate disc, the vesicle then splitting open, or formed by divisions in an originally uniseriate filament, again forming a vesicle which splits open or forms the monostromatic frond directly. The correct generic name for this assemblage of species is *Ulvaria* Ruprecht (1851); the genus originally included the two species *Ulvaria splendens* Ruprecht (1851) and *Ulvaria fusca* (= *Ulva fusca* Postels & Ruprecht, 1840). According to Papenfuss (1960) 'authentic specimens of *U. splendens* and *U. fusca* were examined by Wittrock (1866) who found them to be monostromatic.' Papenfuss (loc. cit.) reported that he borrowed a fragment of one of Postels and Ruprecht's original specimens of *U. splendens* from the Leningrad Herbarium and he confirmed the observation of Wittrock that the plant was monostromatic.

Rosenvinge (1893) regarded *Ulvaria* (*Monostroma*) *splendens* as a variety of *U.* (*Monostroma*) *fusca* and in this he has been followed by a number of authors.

The genus *Monostroma* was created by Thuret (1854) to take two species with the flat monostromatic thallus, *Ulva oxycocca* Kützing as *Monostroma oxycoccum* (Kützing) Thuret (= *M. oxyspermum* (Kützing) Doty, 1947) and *Ulva bullosa* Roth as *Monostroma bullosum* (Roth) Thuret. Experimental culture work since 1960 has shown that *Ulvaria fusca* (= *Ulvaria splendens*) has a life history involving an alternation of isomorphic gametophyte and sporophyte generations, *Monostroma oxyspermum* has an asexual zoospore cycle and *Monostroma bullosum* has a heteromorphic alternation involving a unicellular cyst as the sporophyte phase. A number of authors have placed considerable emphasis on this difference in life history and on this basis have distinguished several genera within the complex. Thus Gayral (1964) proposed three genera, *Ulvaria*, *Ulvopsis* and *Monostroma* with isomorphic, asexual and heteromorphic life histories respectively: Vinogradova (1969) also retained three genera, *Ulvaria*, *Gayralia* and *Monostroma* with similar differences in the life histories. Bliding (1968) separated the complex into three genera, *Ulvaria*, *Monostroma* and *Kornmannia*, the last mainly distinguished by its lack of pyrenoids and by the mode of germination of the rarely produced swarmers. Bliding (loc. cit.) included *Monostroma oxyspermum* with its asexual life history along with *Ulvaria fusca* (= *U. obscura* (Kütz.) Gayral var. *blyttii* (Areschoug) Bliding) which has an isomorphic alternation of generations, a move not approved of by Gayral (1971). It seems unwise to separate genera on the basis of life history unless other distinct characters are associated with it and for the genus *Monostroma* as a whole nine different life histories have been distinguished by Hirose & Yoshida (1964). Also it has been shown (Perrot, 1972) that both isomorphic and heteromorphic life histories can occur for a single species, *Ulothrix flacca*, under different environmental conditions. The difficulties associated with distinguishing taxa within the whole complex of species included in the genus *Monostroma* have been emphasized by van den Hoek (1966).

Tatewaki (1972) distinguished six groupings of species within the complex on the basis of the mode of development of the leafy frond from the zoospore and the method of spore dispersal, as well as on the type of life history. Hori (1972) has eight groupings based on the ultrastructure of the pyrenoid.

When all of the available characters for separation of taxa are considered, no obvious separation on discrete groups of characters appears. The present author therefore prefers to retain one genus for the members of this complex of species with a monostromatic flat thallus until such time as agreement is reached on the number of genera to be recognised and the groups of characters on which to separate them. As the name *Monostroma* has been in general use for so long, it has seemed best to retain it here, even though it is illegitimate, rather than make name changes that may have a limited life.

The genus belongs to cooler climates and is widely distributed in the northern hemisphere, and is known also from the southern hemisphere.

KEY TO SPECIES

1 Frond dull green, turning dark brown to black on drying; cells unordered; chloroplast with more than 1 pyrenoid, up to 6–8; chloroplast often appearing double in surface view with a plate on each side of outer wall . . . . *M. obscurum*
   Frond light green, remaining green on drying; cells in rows or groups; chloroplast with a single pyrenoid   2
2 Texture of thallus gelatinous; cells isolated or arranged in groups of 2 or 4 separated by thick walls; chloroplast cup-shaped . . . . *M. oxyspermum*
   Texture of thallus soft but not gelatinous   3
3 Thallus deeply divided into segments, sometimes almost to base, edges smooth; cells in rows separated by 'veins' of thickening and in groups of 4; basal cells elongated, tapering at both ends . . . . *M. grevillei*
   Margin of thallus very undulate, sometimes divided into broad lobes; cells closely set and angular, in groups of 2–8 cells; chloroplast occupying only centre of cell . . . . *M. undulatum*

Monostroma grevillei (Thuret) Wittrock (1866), p. 57.   Fig. 17; Pl. 1.

Type: PC. Finistère, France.

Enteromorpha grevillei Thuret (1854), p. 25.
Ulva grevillei (Thuret) Le Jolis (1863), p. 37.
Ulvopsis grevillei (Thuret) Gayral (1964), p. 2151.

Adult gametophyte an erect monostromatic foliose lamina, 5–10 cm long, split into segments sometimes almost to base, arising from a filamentous basal system up to 2 mm diameter; prostrate system often becoming discoid by pseudoparenchyma formation; cells of lamina graded in size and shape from distal end to base; distal cells rounded, 5–20 μm diameter, sometimes in groups of 4, cells with 1 pyrenoid; cells of mid-thallus elongated, rectangular, polygonal, or quadrangular, 10–25 μm long, sometimes separated into groups by thickenings giving the appearance of veins; cells with 1 pyrenoid; basal cells very elongate with tapered ends, up to 40 × 5–10 μm,

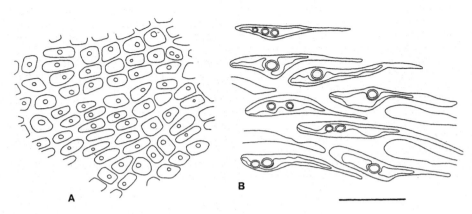

Fig. 17   *Monostroma grevillei*. Adult gametophyte
A. Surface view of vegetative cells (probably mid region). B. Surface view of basal vegetative cells, with tapered ends. Bar= 30 μm.

sometimes with more than 1 pyrenoid. Sporophyte a unicellular *Codiolum*-phase, sometimes spherical, up to 40 μm diameter, lacking rhizoid, sometimes with cell wall extended to form a single or branched rhizoid-like structure; sporophytes with 'rhizoid' may be spherical or elongate, overall dimensions then 100 × 40 μm; chloroplast parietal, reticulate or perforated and plate-like.

Reproduction by biflagellate gametes and quadriflagellate zoospores; plants dioecious; gametes formed in distal cells progressively downwards until whole plant consumed; gametes anisogamous, micro-gametes 5.1–6.0 (mean 5.6) × 2.1–3.0 (mean 2.7) μm, macro-gametes 6.1–7.3 (mean 6.7) × 2.4–4.5 (mean 3.8) μm; zoospores 10.3 × 4.6 μm released either by rupture of sporophyte or by production of hyaline exit tube.

Chromosome no.: not known.

The gametophyte is found growing in rock pools from the upper to the lower littoral zone, on a variety of other algae and on mollusc shells, also in the shallow sublittoral on a variety of substrates including marine angiosperms; the plants occur in both sheltered and exposed sites; the sporophyte is a facultative borer in calcareous substrates.

It is likely that *Monostroma obscurum* and *M. oxyspermum* have been confused with *M. grevillei* in the literature. The British distribution records quoted are therefore from observations confirmed by Dr M. Wilkinson and the author.

Kent, Dorset, Cornwall, Channel Isles, Scilly Isles, Anglesey, Isle of Man, Ayrshire, Renfrewshire, Dunbartonshire, Argyllshire, Fife, West Lothian, Midlothian, East Lothian, Northumberland, Durham.

Spitzbergen and Scandinavia south to France, Baltic Sea, White Sea, Murman Sea, Kamchatka.

Greenland, Iceland, Faroes, Atlantic coast of North America from the Canadian Arctic south to New Jersey; Pacific coast of North America from Alaska south to California, but not recorded for the American Pacific coast by Scagel (1966).

The life history of the plant is an alternation between a foliose gametophyte and a unicellular '*Codiolum*' shell-boring sporophyte. The foliose plant has a marked seasonal behaviour being found on British shores only from late January until early June, but most commonly from February to April and is usually fertile from March onwards. In the Arctic regions the season may be extended until August. It is not known whether or not the prostrate portion of the gametophyte from which the foliose plant arises can perennate from year to year. Since the sporophyte is morphologically indistinguishable from those of *Gomontia polyrhiza* and *Eugomontia sacculata* it is not possible to describe its seasonal behaviour. In rock pools of the upper littoral zone where *M. grevillei* is most abundant shell-boring *Codiolum*-phases have been found throughout the whole year, but at least some of these may belong to other species named above. Kornmann (1962) found that in culture the sack did not upwell from the disc until it was exposed to a temperature of 6°C. It is therefore possible that the prostrate system is an overwintering stage, the sack upwelling when the cold requirement has been fulfilled. *Monostroma grevillei* lives in rock pools appearing first in pools at the top of the littoral zone, but gradually appearing, as the spring progresses, lower down the shore to about the *Fucus vesiculosus/F. serratus* boundary. The plants quickly become reproductive so that they progressively disappear during the spring, also working down the shore. This gives the appearance of a migration downwards. Also, new plants appear in particular pools very shortly after the first wave of plants becomes reproductive, suggesting that there is no obligate alternation of phases in the life history.

The foliose plants rarely reach a length greater than 10 cm, but in very sheltered situations and in strong freshwater inflows, plants up to 100 cm long have been found (Wilkinson, unpublished). The sack normally splits open before the plant is 1 cm in length, but in very sheltered conditions plants have reached a length of 10 cm and become reproductive without the sack splitting.

The prostrate system is usually not clearly visible, but can be up to 2 mm in diameter. It is basically filamentous in nature, but this is obscured to a greater or lesser extent by formation of pseudoparenchyma so that the base is usually discoid according to one of the following three possible mature forms:

1 A disc of rounded cells each with one pyrenoid, about 5 μm in diameter, with filaments so closely compacted and without free filament ends that a compact disc is formed, the disc being initially monostromatic, but finally polystromatic almost to its edges so that it would appear as a thick lens in vertical section.

2 A disc of more clearly filamentous form, developing initially as completely free filaments of elongated cells, about 5 × 10 μm, each with a parietal chloroplast and one pyrenoid, the filaments coalescing later to produce a disc up to 2 mm in diameter consisting of a polystromatic central area of rounded cells 5 μm in diameter and a monostromatic peripheral area of elongated cells, 5 × 10 μm, clearly filamentous and radiating from the central area, with free filament ends.

3 An initially filamentous growth which soon coalesces at its centre to produce a lens-shaped polystromatic disc, resembling type 1 above, from which radiate completely free filaments consisting of very elongated cells, 5 μm in diameter, but up to 100 μm long, only slightly branched and with negligible cell contents, giving the appearance of a rhizoidal growth anchoring the disc.

Although the absence of true rhizoids is a fundamental feature of *M. grevillei*, distinguishing it from *M. oxyspermum*, Wilkinson (unpublished) has found that the basal ends of torn plants of *M. grevillei* can produce rhizoids. This could complicate identification of *M. grevillei* in free-floating populations such as have been reported by Bliding (1968) if the populations contained cast-up plants.

Monostroma obscurum (Kützing) J. Agardh (1882–83), p. 111.   Fig. 18

Type: L. Biarritz, France.

Ulva obscura Kützing (1843), p. 296.
Ulvaria obscura (Kützing) Gayral (1964), p. 2151.

Thallus flat, soft, membranous irregularly split, rarely to base, up to 30 (–40) cm long turning brown on drying, not adhering to paper, outline irregular often with wavy margin, tapering to a stipe; holdfast reinforced by rhizoids arising from cells above holdfast; cells of lamina mostly unordered with occasional indications of rows; cells closely packed, quadrate, rectangular or slightly polygonal, rarely rounded, 18–28 × 10–20 μm, walls gelatinous; lamina up to 100 μm thick in rhizoidal region, thickness above variable depending on position in lamina, 28–60 μm in central lamina, 16–25 μm in marginal zone of upper lamina: chloroplast roughly barrel-shaped often appearing double in surface view due to thickenings with one plate at each outer wall, plates linked through cell; pyrenoids usually more than 1 up to 6–8.

Reproduction by biflagellate gametes 5.3–5.4 × 3.1–3.2 μm and by quadriflagellate zoospores 9.1 × 4.6 μm.

Chromosome no.: n = 9 Tatewaki (1972) (as *M. fuscum*), Dube (1967).

Occurs in the littoral and sublittoral regions of the shore, mainly in pools in the upper littoral, attached to other algae and to shells and stones, often later free floating;

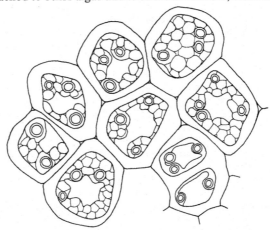

Fig. 18   *Monostroma obscurum*
Surface view of vegetative cells. Bar= 20 μm.

the plants also occur in salt marshes; on the open coast they belong to well illuminated rather sheltered places.

Not often recorded from the British Isles in recent times: Shetland Isles, Durham, Glamorgan, Northumberland, Yorkshire, Berwickshire, Fife, Kent.
Northern Norway south to the Bay of Biscay, Black Sea, Mediterranean.
British Columbia, Oregon and Washington on the Pacific coast of North America, Canadian Arctic coasts; Labrador, Quebec Province on Atlantic coast of North America, Greenland; Japan.

The plants may be present throughout the year, but have been recorded in Britain only from spring to late summer; fruiting season not known. In northern Norway the plants are fertile in July and August and in southern Norway in September and October (Bliding, 1968). In Washington, USA, mature reproductive plants have been found at all seasons of the year. For this area Dube (1967) records a periodicity of release of gametes and zoospores occurring every 8–10 days in July and August, but at intervals in excess of 12 days in December. The life history is an isomorphic alternation of foliose gametophytic and sporophytic generations; gametes may develop parthenogenetically.

A number of varieties of *Monostroma obscurum* (= *M. fusca*), var. *obscurum*, var. *splendens* (Ruprecht) Rosenvinge and var. *blyttii* (Areschoug) Setchell & Gardner, have been described, differing mainly in the thickness of the thallus and in its colour. Setchell & Gardner (1920*a*) and Bliding (1968) have been doubtful of the validity of these varieties based on characters of this kind. Bliding (1968) recorded var. *obscurum* as occupying more southerly latitudes than var. *blyttii*, suggesting that the differences between them may be due to environmental conditions. He also commented that the thickness of the thallus given for var. *obscurum* (var. *fusca*) and for var. *splendens* can be found in different parts of the same plant of var. *blyttii*. As there is no experimental evidence on the genetic status of any of the varieties of *Monostroma obscurum* and the distinguishing character of thallus thickness is difficult to apply, they have here all been included in the species *Monostroma obscurum*.

Monostroma oxyspermum (Kützing) Doty (1947), p. 12.

Type: L. Baltic Sea.

Ulva oxysperma Kützing (1843), p. 296.
Monostroma crepidinum Farlow (1881), p. 42.

Thallus lamina 4–20 cm or more long, deeply divided into lobes of varying width; plants usually attached but may become free-floating and are then of very large dimensions; texture very soft and gelatinous; cells in distinct rows sometimes over large areas, except in basal region; basal cells drawn out into thead-like portions up to 100 μm long; cell shape, except at base, variable, often rounded; cells isolated or arranged in groups of 2–4 separated by thick walls; cells 7–10 (–20) μm long in surface view; chloroplast cup-shaped; pyrenoid single, occasionally 2–3.
Reproduction by biflagellate zoospores liberated by decomposition of fertile tissues of thallus; zoospores 8.4–9.0 × 3.4–4.2 μm.

Chromosome no.: n (?) = 8–10 Tatewaki (1972).

Plants grow at first attached to rock surfaces, to other algae or to shells or small stones throughout the littoral region and the sublittoral fringe and in lagoons, later often forming free-floating masses; the plants occur in rather sheltered places and on the edges of salt marshes where there is overflow or seepage of freshwater to give brackish conditions or fluctuating salinities; the plants can tolerate a range of salinity from about 7–30‰.

Generally common around the British Isles: Kent, Sussex, Dorset, Devon, Cornwall, Anglesey, Isle of Man, Argyllshire, Renfrewshire, Dunbar, Fife, Co. Mayo, Durham.

Norway, Denmark, Sweden, France, Portugal, Baltic Sea, Black Sea, Mediterranean Southern British Columbia to southern California on the Pacific coast of North America; Newfoundland south to Florida and South Carolina on the Atlantic coast of North America; Japan.

Probably present all the year round, but the reproductive season not known for Britain. Plants are fertile from April to November in Sweden (Bliding, 1968). As far as is known the life history involves only an asexual sequence of generations.

This is a very variable species and a number of varieties and forms have been described (see Bliding, 1968). The experimental work of Kornmann & Sahling (1962) suggests that the extreme variability of the plants of the species when grown under different conditions, can cover the forms described in the different species. The mucilaginous nature of the thallus, supposed to be a character of the species, varies with salinity; under higher salinities it is hardly at all gelatinous, but at lower salinities may be markedly so (Bliding, 1968).

There are three records of *Monostroma crepidinum* Farlow for the British Isles all dating from the last century (Batters, 1902). The species was described by Farlow (1881, p. 42) as 'Fronds delicate, light green, one to three inches long, flabellately orbicular, split to the base, segments obovate .018–.036 mm thick, cells roundish–angular, intercellular substance prominent'. Farlow himself realised that it was very close to *M. orbiculatum* Thuret (1854) differing only in the thickness of the thallus, *M. orbiculatum* being 0.032–0.040 mm (*M. orbiculatum* = *M. wittrockii* Bornet). He examined a Thuret specimen of *M. orbiculatum* and found the measurements to be the same as in his new species. Bliding (1968) gave *M. orbiculatum* Thuret as a synonym of *Ulvaria oxysperma* var. *orbiculatum* (Thuret) Bliding. *M. wittrockii* Bornet in Bornet & Thuret (1880) is the basionym for Bliding's *Ulvaria oxysperma* var. *oxysperma* forma *wittrockii* (Bornet) Bliding (Bliding, 1968). *M. crepidinum* Farlow therefore appears to belong within the *Monostroma oxyspermum* aggregate and is treated as such here. British records of *M. orbiculatum* Thuret and *M. wittrockii* Bornet are treated in the same way.

Monostroma undulatum Wittrock (1866), p. 46, pl. III, fig. 9.

Type: S. Sweden.

M. pulchrum Yendo (1917), p. 186 nom. illeg., non *M. pulchrum* Farlow (1881), p. 41.

Leafy frond light green, soft, flaccid, up to 15–25 cm long, sometimes divided into lobes, margin very undulate; frond 40–50 μm thick, developing by divisions in a primarily uniseriate filament; cells closely set, angular, in groups of 2–8 cells, cells 20–24 μm long, oval in cross section; chloroplast appearing as a central band in cell. Unicellular thick-walled cysts pear-shaped or obovoid.

Reproduction by quadriflagellate zoospores 5–12 × 3–6 μm, lacking eyespot, formed in cells of leafy frond and in thick-walled cysts; zoospores of leafy frond released by disintegration of fertile region; zoospores from cysts released enclosed in hyaline sack.

Chromosome no.: both leafy frond and cyst gave a count of ca. 13 (Tatewaki, 1972).

Found growing on stones or epiphytic in the upper littoral on open coasts.

The only British record is from the Orkney Islands (Ronsay; Mrs Traill about 1837 in herb. J. H. Pollexfen, see Batters, 1902). Not recorded since then.
Recorded for Helgoland by Kornmann & Sahling (1962), Norway.
Faroes, Greenland south to Massachusetts (?); Japan.

In Helgoland plants appear in March–April and diminish in May–June. There is no information for plants in the British Isles. The life history is a heteromorphic alternation of generations; both generations are asexual as shown by the culture work of Tatewaki (1972) and Kornmann & Sahling (1962). Tokida (1954) in Japan reported biflagellate swarmers formed in the same individual leafy fronds as quadriflagellate zoospores 'often found to be fused to each other as if they were conjugating gametes'. In Tokida's opinion these represented not conjugating gametes, but incompletely divided swarmers 'because they are fused always laterally at the posterior parts and the true conjugation movement could never be observed'.

The relationship between *M. undulatum* Wittrock, var. *undulatum*, var. *farlowii* Foslie and *M. pulchrum* Yendo (non Farlow) seems a little uncertain. Kornmann & Sahling (1962) thought that, from the observations of Yamada & Saito (1938) on *M. pulchrum*, that it was the same as the Helgoland plant they were calling *M. undulatum*. They also assumed that it was the same as *M. undulatum* var. *farlowii* described by Tokida (1954). Tokida gave *M. pulchrum* Yendo (non Farlow) as a synonym of *M. undulatum* var. *farlowii*, but he does not comment on Foslie's var. *farlowii* of *M. undulatum*. Hirose & Yoshida (1964) also record *M. pulchrum* as a synonym of *M. undulatum* Wittrock and record *M. pulchrum* Farlow as a separate entity.

ACROSIPHONIACEAE Jónsson

Acrosiphoniaceae Jónsson (1959b), p. 1567.

Codiolaceae Den Hartog (1959), p. 111.

Thallus differentiated into prostrate and upright system, simple or branched filaments with uninucleate or multinucleate cells; branched thalli with conspicuous apical cells and with laterals often differentiated into rhizoidal or more complex hooked branches; nuclear and cell divisions synchronised; chloroplast an archaeoplast, i.e. a perforated net containing one or more polypyramidal pyrenoids.

Reproduction by biflagellate gametes formed in cells of filamentous gametophyte, or by quadriflagellate zoospores formed in cells of unbranched filaments or in unicellular *Codiolum/Chlorochytrium*-like sporophytes.

The family Acrosiphoniaceae was established by Jónsson (1959) to take the genera *Acrosiphonia*, *Spongomorpha* and *Urospora*, originally classified in the Cladophoraceae, but which differed from the other members of this family in a number of characters, the most important of which, perhaps, being the heteromorphic life history involving a unicellular sporophyte or *Codiolum*-phase. Other important characters are the presence of a parietal archaeoplastid in the form of a perforated net, containing polypyramidal pyrenoids and a synchronisation of nuclear and cell divisions. The family also differs from the Cladophoraceae in the breakdown products of its soluble polysaccharides (O'Donnell & Percival, 1959) and in the absence of crystalline cellulose I from the cell walls (Nicolai & Preston, 1952, 1959). The chemistry of the three genera brings them nearer to the Ulotrichales than to the Cladophorales (see also discussion under Cladophorales, p. 136).

The family Codiolaceae was introduced by Den Hartog (1959) to take the genus *Codiolum* and species of *Urospora* said to have this type of heteromorphic life history. The name Codiolaceae is antedated by Acrosiphoniaceae (Jónsson, 1959*b*) based on *Acrosiphonia*, but including *Urospora*, one character being the heteromorphic life history. The name Codiolaceae is therefore illegitimate (Silva, 1980). It should also be remarked that several different life histories have been described for species of *Acrosiphonia*, not all of them including a *Codiolum*-phase.

Acrosiphonia and *Spongomorpha* were originally separated on the uni or multi-nucleate conditions of the cells (see Scagel, 1966). There seems little justification for this and, here, Parke & Dixon (1968) have been followed in including *Acrosiphonia* in *Spongomorpha*.

SPONGOMORPHA Kützing

SPONGOMORPHA Kützing (1843), p. 273.

Type species: *Spongomorpha uncialis* (Müller) Kützing (= *Spongomorpha aeruginosa* (Linnaeus) van den Hoek).

Acrosiphonia J. Agardh (1846), p. 82.

Thalli with both prostrate and erect systems of branched filaments; prostrate filaments free or compacted into an irregular flat plate; erect filaments broadening from base to apex with conspicuous apical cells with bluntly rounded tips; axes with laterals differentiated into terminal erect, curled, sometimes spinous, and rhizoidal branches; branches at first spreading, later becoming bound together so that axes form ropes; filaments with apical segmentation, later intercalary; vegetative cells uninucleate or multinucleate; pyrenoids many, embedded in a single parietal, reticulate chloroplast; cells with great capacity for regeneration. Sporophyte when present, unicellular, *Codiolum*-like, developing from zygote.

Reproduction by biflagellate gametes formed in isolated cells of erect filaments, or in rows of cells, 25–30 per cell; sometimes by biflagellate neutral spores; sometimes by quadriflagellate zoospores formed in unicellular sporophyte. Vegetative reproduction by fragmentation and regeneration. Several different life histories have been described for the genus; so far the position of meiosis has not been established.

The plants of the genus *Spongomorpha* are easy to recognise by the form of the axes, the cells of which become broader from the bases to their apices and which have very conspicuous apical cells; also by the differentiation of the axes which form rhizoidal branches in the lower part and sometimes curled and hooked or spinous branches in the central part of the thallus. Many species have been described, the majority of which have now been recognised as growth forms of a few species. Two species are recognised in the British Isles.

KEY TO SPECIES

Filaments 40–120 μm diameter, increasing towards apex of plant: distinct hooked and spinous branches present except in very young plants: cells multinucleate .... *S. arcta*
Filaments mostly 30 (–35) μm diameter, scarcely increasing towards plant apex: sometimes with plentiful hooked, but never spinous branches: cells uninucleate .... *S. aeruginosa*

Spongomorpha aeruginosa (Linnaeus) van den Hoek (1963), p. 225.

Lectotype: OXF (*Conferva marina capillacea brevis viridissima mollis* Dillenius (1741), p. 23; see van den Hoek, 1963, p. 225). Anglesey.

Conferva aeruginosa Linnaeus (1753), p. 1165.
Cladophora lanosa Harvey (1846), pl. VI.
Chlorochytrium inclusum Kjellman (1883), p. 320.
Codiolum petrocelidis Kuckuck (1894), p. 259.

Plants bright green, fading to dull yellow and eventually whitish, 1–3 (–4–5) cm long; thallus of filaments bound to some degree into ropes; growth at first apical, later intercalary; branching dense or less so, lower branches lying at wide angle due to changes in wall thickening; filaments usually 30 μm diameter, sometimes 35 μm, apices blunt, straight or hooked; rhizoids downward growing, slender, 20–30 μm diameter, often with re-attachment and stolon formation; cells above up to 12 times as long as broad, thin walled, lower down following intercalary division $\frac{1}{2}$–$2\frac{1}{2}$ times as long as broad or shorter than broad, with thicker walls; rhizoid cells usually 6–12 times as long as broad with dense tips; stolon or re-attachment plates with dense contents and thick walls; chloroplast parietal, perforated with several pyrenoids, nucleus single.

Reproduction by biflagellate gametes and quadriflagellate zoospores. Gametangia from isolated cells or rows of cells forming 25–30 gametes per cell; gametes pyriform with single chloroplast, pyrenoid and reddish-brown stigma, escaping *en mass* through simple pore, isogamous or anisogamous; plants monoecious. Zoospores formed in unicellular sporophyte developing from zygote, identifiable as *Chlorochytrium inclusum* Kjellm. or *Codiolum petrocelidis* depending on host plant.

Chromosome no.: 2n = 12   Patel (1961) as *Spongomorpha lanosa* Kütz.
            2n = 30 ? Föyn (1929) as *S. lanosa* (S. *uncialis* Kütz.) reported
                  by Godward (1966).
            n = c. 6 Jónsson (1962).

The plants are found growing epiphytically on *Polyides* and other red algae, also on brown algae, near the lowest littoral levels and in the sublittoral down to a depth of several metres. Sporophyte endophytic in *Petrocelis* (= *Codiolum petrocelidis*) or in *Polyides*, *Dilsea* and other algae (= *Chlorochytrium inclusum*).

The species is fairly common and locally abundant all round the coast of the British Isles.

European Atlantic coast from Scandinavia south to France.

Atlantic coast of North America from the Canadian Arctic south to Connecticut, Greenland.

The alga referred to *Codiolum petrocelidis* on the Pacific coast of North America has been found to be part of the life history of *Spongomorpha coalita* (Fan, 1957, 1959; Hollenberg, 1957, 1958).

Plants of *Spongomorpha aeruginosa* can be found on the shore between mid-March and mid-August, but are only common from mid-April, and can be found fertile from March onwards. The sporophyte is present and fertile mainly during the winter months but can be found fertile during the summer too. The life history is a heteromorphic alternation between an erect filamentous gametophyte and an endophytic *Codiolum/Chlorochytrium*-like sporophyte. The plants have a certain capacity for vegetative reproduction through the re-attachment of detached branches.

Plants can become bound into ropes by their curled branches, but there is no marked age-related change in morphology comparable with that shown by *Spongomorpha arcta*.

Spongomorpha arcta (Dillwyn) Kützing (1849), p. 417.   Fig. 19; Pl. 2

Holotype: BM-K? (Material collected by Miss Hutchins). Bantry Bay, Ireland.

Conferva arcta Dillwyn (1809), p. 67.
Acrosiphonia arcta (Dillwyn) J. Agardh (1848), p. 12.
Conferva centralis Lyngbye (1819), pl. 56.
Acrosiphonia centralis (Lyngbye) Kjellman (1893), p. 73.

Gametophyte of erect branched filamentous axes, at first free, later bound together, up to 10–15 cm long; colour very bright green, darkening with age; branching primarily monopodial, becoming dense and more irregular with age; branches arising in upright position, later lying at wide angle due to change in cell wall at base of branch; segmentation originally apical, becoming intercalary; branches of several types, originally blunt, either straight or hooked, later becoming spinous at tips as growth ceases, first in lower plant, later above; cells 40–120 μm diameter, broadest near apex, upper cells 10–20 times as long as broad, lower cells 1–2$\frac{1}{2}$ times as long as broad, or shorter than broad; descending rhizoids present throughout plant, slender, 40–60 μm diameter, of cells 2–6 (–12) times as long as broad with pale chloroplast except in dense tips; tips may re-attach and form thick-walled storage cells; rhizoids and hooked branches often binding filaments into ropes; cells multinucleate with many pyrenoids embedded in single reticulate parietal chloroplast; nuclear and cell divisions synchronised; sporophyte reported to be a unicellular *Codiolum*-like structure.

Reproduction by biflagellate gametes and sometimes by quadriflagellate zoospores; gametes formed in unmodified vegetative cells in rows in erect filaments, 6–8 μm long, pyriform with chloroplast, pyrenoid and red-brown stigma, escaping through lateral

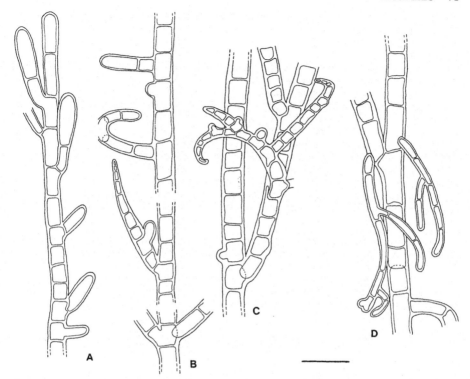

Fig. 19   *Spongomorpha arcta*
A–D. Portions of erect filaments. Bar= 200 μm.

pore with hinged lid; gametes isogamous, plants monoecious. Vegetative reproduction by fragmentation and regeneration from prostrate system, also by new prostrate systems formed by rhizoids on contact with substrate.

Chromosome no.: n = 5 Jónsson (1962).

The species occurs in mid- or lower littoral rock pools or on rock surfaces exposed by the tide on both sheltered and exposed coasts.

It is generally common round the coast of the British Isles.

Finland; Atlantic coast of Europe from Scandinavia south to France.

Atlantic coast of North America from the Canadian Arctic south to New Jersey; North American Pacific coast from the Bering Sea south to Washington; Iceland; Kamchatka.

The species overwinters as a prostrate filamentous system or possibly sometimes as a unicellular sporophyte. The erect system of filaments appears on the shore in February and can be found through to August or mid-September. Reproduction occurs from

March onwards and release of gametes is correlated with the occurrence of low water of spring tides (Archer, 1963). The season is most extended on northern shores. Vegetative reproduction is important ,in the spread of the species and in its overwintering.

The life history of *Spongomorpha arcta* is still a little uncertain. Jónsson (1962) described a heteromorphic life history involving an erect filamentous gametophyte and a *Codiolum*-phase sporophyte for the species *Acrosiphonia spinescens* which he regarded as conspecific with *Acrosiphonia* (= *Spongomorpha*) *arcta*. Later (1964) he found in culture experiments that the two nuclei did not always fuse in the zygote but became separated by a cell wall; the *Codiolum*-phase was not then formed and the zygote developed directly into the erect gametophyte. Kornmann (1965, 1970) described a life history for *Acrosiphonia* (= *Spongomorpha*) *arcta* in which there was only one phase, the erect filamentous gametophyte which developed directly from the zygote. However, he regarded *Acrosiphonia spinescens* as a separate species which he described as having the heteromorphic life history. Since *A. arcta* and *A. spinescens* are morphologically indistinguishable, it becomes a question as to how the species should be defined.

Young plants of *Spongomorpha arcta* are open and radiate, but later the filaments become roped together by the hooked branches and intertwining rhizoids, the ropes eventually becoming shorter and coarser following loss of the distal parts following reproduction. The changes taking place as a plant develops and matures have in the past been a source of considerable confusion in the delimitation of the species, the different stages having been given different specific names.

ULVACEAE Lamouroux ex Dumortier

ULVACEAE Lamouroux ex Dumortier (1822), pp. 72, 102.

Ulvacées Lamouroux (1813), p. 275.
Enteromorphaceae Kützing (1843), p. 300 (as Enteromorpheae).
Phycoseridaceae Kützing (1843), p. 296 (as Phycoserideae).
Capsosiphonaceae Chapman (1956), p. 429.
Percursariaceae Bliding (1969), p. 618.

Thallus parenchymatous of three differing forms, a biseriate filament, a simple or branched tube with a monostromatic wall, or a distromatic entire or lobed blade; thallus attached by rhizoids often associated to form a flat disc; setae absent; chloroplast a parietal plate or stellate with 1–12 pyrenoids.

Reproduction by quadri- or biflagellate zoospores or biflagellate gametes formed in unmodified cells of the vegetative thallus; gametes isogamous or anisogamous.

The family Ulvaceae is presented here in the traditional sense of including parenchymatous plants of varied morphology, all of which are derived from uniseriate filaments by divisions of the cells in two or three planes. Included are mainly unbranched or branched hollow tubes or flat leaf-like structures. The thalli, except for the biseriate filament of *Percursaria*, are fundamentally tubular being formed by transverse and longitudinal radial divisions in the primary filament: longitudinal tangential divisions are limited to branch formation and solid parenchymatous structures are not formed. The genus *Capsosiphon*, despite the anomalies shown, has been included in the family and *Monostroma* excluded for reasons already discussed (see p. 58).

The Ulvaceae has many features in common with the Chaetophoraceae, e.g. details of the sporangium structure and development, mode of release of swarmers, life history patterns, structure of the pyrenoids and absence of plasmodesmata (O'Kelly 1980). Siphonoxanthin occurs in species of both families. The two families have been kept together in the Ulotrichales for the time being.

CAPSOSIPHON Gobi

Capsosiphon Gobi (1879), p. 88.

Type species: *Capsosiphon aureolum* Gobi (1879), p. 88.

Thallus tubular, hollow or compressed, with a wall of a single cell layer, attached by a disc; cells clustered in groups of 2 or 4, with or without distinct common parent cell sheaths, arranged in distinct longitudinal rows, sometimes spirally disposed, loosely held together by gelatinous walls; branching of thallus by separation of cell rows; chloroplast parietal with a single pyrenoid. Reproduction by quadriflagellate or biflagellate zoospores and/or biflagellate isogametes; zygotes form cysts which release quadriflagellate zoospores; reproduction also by aplanospores.

For discussion on the genus *Capsosiphon* see the general discussion on the Ulotrichales (p. 42).

Only one species in Britain.

Capsosiphon fulvescens (C. Agardh) Setchell & Gardner (1920), p. 280.

Lectotype: LD (No. 13358 in Herb. Alg. Agardh.). Landskrona, Sweden.

Ulva fulvescens C. Agardh (1822), p. 420.
Enteromorpha aureola (C. Agardh) Kützing (1849), p. 481.
Capsosiphon aureolum (C. Agardh) Gobi (1879), p. 88.

Plants filamentous, up to 20 or more cm long, 3–6 mm broad, becoming tubular, hollow, often spirally twisted, attached by colourless or pale green rhizoids or by a disc; occasionally a few branches similar to axes formed by separation of cell rows; plants golden brown or green, gelatinous, cylindrical or compressed; cells in groups of 2 or 4 enclosed in the mucilage sheath of the mother cell; cells rounded, oval or quadrate with rounded corners, (4–) 5–10 × 6–12 (–14) μm; cell walls thick and mucilaginous; chloroplast parietal with single large pyrenoid.

Reproduction, in any area of thallus, by quadriflagellate or biflagellate zoospores formed in unmodified vegetative cells, 8–16 (–24) per cell, released in a membranous vesicle; zoospores 7.5–10 × 3.5–8.0 μm with red eye spots and parietal chloroplasts; reproduction also by biflagellate isogametes; zygotes forming cysts releasing 16–32 quadriflagellate zoospores.

Chromosome no.: not known.

Found growing on rocks and boulders, woodwork, sea walls, piers and logs with a covering of silt in the mid- and upper littoral region of the shore at the mouths of rivers and streams and also on open shores with freshwater drainage; found in polluted harbours; often mixed with *Prasiola stipitata, Blidingia* species and *Urospora penicilliformis*.

Occurs all round the coast of the British Isles from Shetland to the south coast in suitable habitats and is probably more common than the records indicate.

European Atlantic coast from Scandinavia to France, Gulf of Finland, Mediterranean, Baltic Sea, Adriatic Sea.

Atlantic coast of North America from Labrador south to New Jersey; Iceland; Japan.

In Britain and on the Atlantic coast of Canada plants appear in the early part of the year and fruit during the summer months. It is reported that in Japan young thalli appear in mid-November, with luxuriant growth in February and then disappear before the end of April; fruiting occurs throughout the growing season (Chihara, 1967; Migita, 1967). In Helgoland the plants are reported as fertile throughout the year (Kornmann & Sahling, 1977).

Only an asexual life history has been reported for *Capsosiphon fulvescens* in Britain and the same is true for plants in Nova Scotia (pers. obs.). Bliding (1963) reported a similar life history, involving only quadriflagellate zoospores, for some plants in Sweden, but he also found the occasional production of biflagellate swarmers by some plants. Iwamoto (1959) reported biflagellate swarmers from plants near Tokyo, which he suspected were gametes, though he saw no fusions. Chihara (1967) described only sexual generations for Japanese plants, but Migita (1967), for plants found near Nagasaki, Japan, found both zoospore and gamete production, the zoospores germinating directly to form filamentous thalli; gametes from different thalli conjugated isogamously to form zygotes which developed into cysts, each forming 16–32 quadriflagellate zoospores, which on germination grew to 'filamentous leafy thalli'. Migita (loc. cit.) also observed parthenogenetic germination of gametes. Yoshida (in Migita, 1972) found cyst stages in cultures of *C. fulvescens* and he stated that the species is closely related to *Monostroma groenlandica*. So far the actual site of meiosis in the life history has not been determined.

The growth rate of the plants in culture is higher in full seawater than at lower salinities and the plants die if kept continously in freshwater; also a salinity approaching that of seawater is required for swarmer release. It looks therefore as though *Capsosiphon* can tolerate periods of immersion in freshwater rather than preferring it as stated by Bliding (1963). The separation of the longitudinal rows of cell groups is more marked at low than at higher salinities (Griffiths, unpublished).

BLIDINGIA Kylin

BLIDINGIA Kylin (1947), p. 8.

Type species: *Blidingia minima* (Nägeli ex Kützing) Kylin (1947), p. 181.

Feldmannodora Chadefaud (1957), p. 668.

Thallus tubular, unbranched or branched, several thalli arising together by raising of central portions of a prostrate monostromatic disc; cells of disc non-rhizoidal; cells of thalli forming a single layer, arranged irregularly or in longitudinal rows; cell length no greater than 9 μm, usually about 7 μm; chloroplast stellate with one pyrenoid.

Reproduction by quadriflagellate zoospores or occasionally by biflagellate swarmers; both spore types are formed in unmodified vegetative cells towards the apex of the thallus; biflagellate swarmers and sexual reproduction not known in Britain. On germination the zoospores produce a germination tube into which the contents of the zoospore pass: an empty basal cell is cut off and the thallus develops from the upper cell.

The genus *Blidingia* was separated from *Enteromorpha*, which it resembles morphologically, by its smaller cell size, the formation of a germ tube, in the development of the zoospore and the formation of a prostrate disc of cells from which the erect thalli arise by raising of areas of cells (Kylin, 1947). Three species of *Blidingia* are described for the British Isles. They differ in cell size and arrangement, in the thickness of the inner cell wall membrane, in the behaviour of the germ tube and in the form of the prostrate disc. *B. chadefaudii* differs only a little in cell size from *B. minima* and the internal wall thickenings, an important diagnostic character in the original description of *B. chadefaudii,* are not always marked according to Kornmann & Sahling (1978). *B. marginata* differs from the other two species of *Blidingia* in having its cells arranged more or less in regular longitudinal rows. However, the regular and irregular arrangements can merge into one another in a single thallus and specific distinction is not always easy.

KEY TO SPECIES

1 Cells arranged in distinct longitudinal rows . . . . *B. marginata*
   Cells irregularly arranged   2
2 Cells 6–7.5 (–9) μm long with thickened internal walls . . . . *B. chadefaudii*
   Cells 4.5–5.5 (–7) μm long, without internal wall thickenings . . . . *B. minima*

Blidingia chadefaudii (Chadefaud) Bliding (1963), p. 30.

Type: PC? Near Roscoff, Finistère, France.

Feldmannodora chadefaudii Chadefaud (1957), p. 653.
Enteromorpha chadefaudii Feldmann (1954), nom. nud.

Thalli usually unbranched, often with small proliferations from base, 2–8 cm long, 2–4 mm broad, narrowing to short stipe, attached by monostromatic disc formed of loosely associated branched filaments, distromatic in centre, sometimes attached also by rudimentary rhizoids; cells initially in regular rows, later irregularly arranged, polygonal to rounded with internal walls often strongly thickened, thickenings directed towards apex of thallus; cells 6–7.5 μm lumen diameter, up to 10 μm in fertile region, cells elongate up to 30 μm or longer; chloroplast stellate with a single large pyrenoid.

Reproduction by quadriflagellate zoospores (biflagellate according to Chadefaud (1957)), 4 per cell, released in a hyaline membrane; zoospores lacking eyespot.

Chromosome no.: unknown

Forming a short turf on wave and spray washed rocks in the upper littoral region; sometimes mixed with *Blidingia minima*, sometimes forming a separate zone (Kornmann & Sahling, 1978).

Not seen by the author, but recorded from the south coast of Britain. The very limited records of this species may be due to confusion with *Blidingia minima*.

Norway, Helgoland, France, Mediterranean.

Recorded as present in spring, summer and autumn. The species may be present all the year round but information is incomplete. The fruiting period is also uncertain.

The life history appears to be a succession of asexual generations: the function of the biflagellate swarmers is not known.

The plants seem to vary in the degree of thickening of the inner cell walls of the thallus. Chadefaud (1957) described and illustrated them as exceedingly thick and regarded this as a key character separating the species from *B. minima*. Kornmann & Sahling (1978), however, found no such thickenings in plants in Helgoland which they identified as *B. chadefaudii*. Bliding (1963) found that the thickness of the wall decreased in plants kept continuously in water through several generations in culture, though they remained somewhat thickened. The thalli are sometimes host to *Chlorochytrium moorei*, especially during the summer months.

Enteromorpha chadefaudii was named by Feldmann (1954) for plants he collected in 1947 and 1948 at Point de Primel, north east of the Bay of Morlaix in Brittany, France. The name was not accompanied by either a diagnosis or an illustration. The name *E. chadefaudii* is therefore a nomen nudum (see Chadefaud, 1957). Chadefaud created a new genus *Feldmannodora* to take the entity, but later Bliding (1963) decided, on the basis of his studies, that the entity rightly belonged in the genus *Blidingia* and transferred it there under the name *Blidingia chadefaudii* (Chadefaud) Bliding.

Blidingia marginata (J. Agardh) P. Dangeard (1958), p. 347.   Fig. 20A

Lectotype: LD (No. 14161 Herb. Alg. Agardh). Nice, France.

Enteromorpha marginata J. Agardh (1842), p. 16.
Enteromorpha micrococca Kützing (1856), pl. 30, fig. 2.

Thalli 1–5 (–20) cm long forming soft masses over substrate; filaments attached to compact prostrate disc; thalli cylindrical when small, later becoming compressed, with occasional branches, highly branched under low salinity conditions; cells in sometimes indistinct, longitudinal rows, alignment accentuating margins in surface view; cells rounded or quadrate 7.5 (–8) × (4–) 6 (–8) μm; chloroplast stellate, pyrenoid single.

Reproduction by quadriflagellate zoospores; biflagellate swarmers formed occasionally.

Chromosome no.: unknown

Found growing in the upper littoral and supralittoral zones, sometimes around bases of higher plants, on vertical or steeply sloping rock faces and on isolated boulders and stones; also on salt marshes; sometimes mixed with *Blidingia minima*; able to remain dry for long periods under unfavourable moisture conditions.

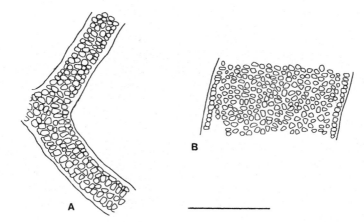

Fig. 20   *Blidingia* spp.
A. *B. marginata*. Portion of erect thallus showing regular arrangement of cells. B. *B. minima*. Portion of erect thallus showing irregular arrangement of cells. Bar= 50 μm

Distributed around the whole coast of the British Isles on suitable substrates and penetrating into estuaries.

European Atlantic coast from Norway to Portugal; Mediterranean.

American Atlantic coast from the Canadian Arctic to Connecticut and New Jersey; Bermuda; New Zealand.

The species is present and reproduces throughout the year, though plants become dry and white during dry weather. In culture Kornmann & Sahling (1978) obtained plants 8 mm long, already reproducing, four weeks after settlement. The life history involves a succession of generations producing only zoospores.

The thalli reach a greater length and width under brackish water conditions and become highly branched, branching to several degrees, with pointed branches. This morphological condition was recorded as a variety, *B. marginata* var. *subsalsa* (Kjellman) Scagel by Scagel (1957). It was raised to a subspecies, *B. marginata* subsp. *subsalsa* (Kjellman) Bliding by Bliding (1963). Experimental work (pers. unpublished) suggested that this was a straightforward habitat modification, the condition being produced by growth at low salinities. Kornmann & Sahling (1978), on the basis of their culture experiments using Helgoland material, reported that the characters were inherited and hence proposed the taxon *B. subsalsa* (Kjellman) Kornmann & Sahling.

Blidingia marginata acts as host to *Chlorochytrium moorei*.

The species differs from *B. minima* and *B. chadefaudii* in that no empty cell is cut-off during germination of the zoospores. It differs also in the form of the prostrate disc (see Fig. 0).

Blidingia minima (Nägeli ex Kützing) Kylin (1947), p. 181. Fig. 20B

Holotype: L (938.69...168). Helgoland.

Enteromorpha minima Nägeli ex Kützing (1849), p. 482.

Thalli 1–5 (–24) cm long, unbranched or branched, up to 4 mm broad, attached in groups to a common basal disc; cells mostly irregularly arranged, rounded or polygonal, (3–) 4–7 (–9) μm broad; chloroplast stellate with single pyrenoid.

Reproduction by quadriflagellate zoospores formed in upper cells of thallus, 4–8 per cell; zoospores 5.4 × 4.3 μm.

Chromosome no.: not known.

Found on rocks, boulders and stones in the upper littoral region; also on harbour walls and in estuaries from brackish to almost freshwater regions, occasionally free-floating in sheltered bays, lagoons and estuaries; found on both exposed and sheltered shores; sometimes forming more or less pure stands, sometimes mixed with other *Blidingia* species.

Found round the whole coast of the British Isles.

Occurs from Spitzbergen to the Mediterranean and the Adriatic Sea on the European Atlantic coasts. The world distribution is sometimes given as 'ubiquitous' (e.g. Pringle, 1978), the most northerly observation being Spitzbergen at latitude 79°54' N (Kjellman, 1883) and the most southerly being South Island, New Zealand at latitude 47°S (Chapman, 1976).

Plants can be found reproducing throughout the year, with the best development during the spring and summer months. They are reported to persist through the winter months encased in ice in the Arctic Sea, resuming development when the ice melts (Kjellman, 1883). The life history is a succession of generations producing quadriflagellate zoospores. Biflagellate swarmers have been observed released from plants collected in Dorset, but their function is obscure (Ellis, pers. comm.).

Laboratory culture experiments have shown that if zoospores from open shore unbranched plants are grown in a medium with a salinity of 21‰, 90–100% of them are branched in the way described by Bliding (1963). He described these as *B. minima* var. *ramifera* Bliding. This growth form is found growing in brackish or almost freshwater (Mills, pers. comm.). Pringle (1975) showed that increase in abundance of *Blidingia minima* is associated with brackish water, high light intensity, high air temperature, moderate desiccation, and, possibly, favourable ion ratios. The upper limit of distribution appeared to be associated with unfavourable osmotic conditions, e.g. heavy rain or extreme desiccation; the lower limit with low light intensities. Growth of the plants is favoured by pollution (Sundene, 1953).

ENTEROMORPHA Link in Nees, nom. cons.

ENTEROMORPHA Link in Nees (1820), p. 5.

Type species: *Enteromorpha intestinalis* (Linnaeus) Greville (1830), p. lxvi.

Splanchnon Adanson (1763) Fam. 2, 13.

Solenia (C. Agardh) C. Agardh (1824), p. XXXII, 185.
Ilea Fries (1825), p. 336.
Tubularia Roussel (1806), p. 98.
Fistularia Greville (1824), p. lxvi, 300.
Hydrosolen Martius (1833), p. 10.
Kallonema Dickie (1871), p. 457.
Gemina Chapman (1952), p. 51.

Thallus a hollow tube bounded by a single layer of cells dividing by transverse and radial longitudinal walls, tangential longitudinal divisions in isolated cells giving rise to characteristic branching patterns, or unbranched; attached by rhizoidal branches sometimes cohering to form an irregular disc from which new fronds may arise; no obligate primary disc; chloroplast a parietal plate, lobed or toothed; pyrenoids one to many per cell; chloroplast sometimes rotating with direction of incident light. Reproduction by biflagellate anisogamous gametes and by quadriflagellate and biflagellate zoospores; gametes and zoospores formed in unmodified vegetative cells at distal end of thallus; reproduction may be solely by parthenogenetic gametes or by zoospores.

Within the species of *Enteromorpha*, some of which are distributed from the poles to the tropics and from fully saline to freshwater conditions, it is possible that there are many genetic races adapted to conditions over restricted parts of the ranges. Whether or not such races exist and whether they are interfertile can only be determined by experiment. Bearing this in mind and in view of the demonstrable flexibility of many of the species, it has seemed wise to treat them on a broad basis until such time as the full picture is known. Bliding (1963) in his comprehensive review of the European species of *Enteromorpha*, separated them into groups, each referable to a specific name. Although the groups themselves are fairly readily recognisable by morphological and anatomical characters, it is sometimes extremely difficult to distinguish the individual taxa within them. Studies on the morphological and anatomical characters of British *Enteromorpha*, together with intersterility tests and life history studies, led Silva (1969) to recognise the following species for the British Isles:

E. *torta* (Mert. in Jürg.) Reinb.

E. *ralfsii* Harv.

E. *prolifera* (O. F. Müll.) J. Ag.

E. *flexuosa* (Wulf. ex Roth) J. Ag,

E. *clathrata* (Roth) Grev.

E. *ramulosa* (Sm.) Hook (= E. *crinita* (Roth) J. Ag.)

E. *linza* (L.) J. Ag.

E. *intestinalis* (L.) Link with two subspecies, subsp. *intestinalis* and subsp. *compressa*.

These correspond approximately to the groups of Bliding (1963) with the exceptions that two species, E. *torta* and E. *ralfsii*, were recognised in Bliding's '*torta*' group and two species E. *clathrata* and E. *ramulosa* in his '*clathrata*' group. In one case, that of the '*intestinalis*' group, two species E. *intestinalis* and E. *compressa* were shown by experimental work to be ecotypes of one species and these are therefore treated as two subspecies of E. *intestinalis* (see Silva & Burrows, 1973). More recently Polderman (1975) has suggested that E. *torta* is merely a habitat form of E. *prolifera*. This has been accepted here.

Of the characters traditionally used to distinguish species of *Enteromorpha*, the following have been shown to be the most stable:

1　cell arrangement, this is dependent on planes of cell division
2　the number of pyrenoids per cell, particularly in the basal part of the frond
3　the shape of the chloroplast, if allowance is made for the fact that in some
　　species the chloroplast can move its position in relation to the direction of
　　incident light (Bliding, 1963; Silva, 1969)

Most of the species seem to have a capacity for proliferation. A number of authors have shown that if the thallus of *Enteromorpha* is cut or torn, except in the reproductive region, proliferations can arise from the cut surface. The basal end of a cut piece will give rise to rhizoids and the upper end to vegetative branches (Müller-Stöll, 1952; Burrows, 1959; Lerston & Voth, 1960; Eaton *et al.*, 1966; Silva, 1969). This process can give rise to unusual branching in plants that normally have a very characteristic pattern of branching, or no branching at all. In its extreme form it results in the so-called 'bottle-brush' thalli (Reed & Russell, 1978; Moss & Marsland, 1976). Proliferous branching may also arise by the development of swarmers inside the sporangial cells or by development of latent thick-walled cells in thalli previously subjected to unfavourable conditions. Branching of this proliferous type can be produced by the shock of extreme environmental conditions such as a drop in temperature or high salinity following evaporation in shore pools (Burrows, 1969; Moss & Marsland, 1976; Reed & Russell, 1978). Branching pattern must therefore be used with extreme caution as a specific character unless the cause of the condition is known.

KEY TO SPECIES

1　Diameter of thallus more or less the same from base to apex; cells in three
　　or more longitudinal rows 2
　　Diameter of thallus increasing from base to apex 3
2　3–12 distinct longitudinal rows of cells; thallus with cavity about 12 μm
　　diameter; majority of cells with one pyrenoid, a few with up to 3
　　pyrenoids . . . . *E. prolifera*
　　3–8 distinct longitudinal rows of cells; thallus cavity about 8 μm diameter;
　　majority of cells with more than 1 pyrenoid, usually 2–8 . . . . *E. ralfsii*
3　Cells irregularly arranged in the middle and basal regions of the thallus;
　　small areas, of at least 10–12 cells regularly arranged, not uncommon in
　　the middle region; chloroplast usually hood-shaped and at the apical
　　end of the cell . . . . *E. intestinalis*
　　Cells in the middle region of the thallus in longitudinal rows and usually
　　also in transverse rows; chloroplast not always at the apical end of the
　　cell 4
4　The majority of cells at the base of the thallus with 1 pyrenoid, only a few
　　with 2 or more 5
　　The majority of cells at the base of the thallus with 2–4 pyrenoids, only a
　　few with 1 . . . . *E. flexuosa*
　　The majority of the cells at the base of the thallus with more than 5
　　pyrenoids, usually 8–10 or more 6
5　Thallus highly compressed, unbranched; walls of central cavity closely
　　adpressed, free only at margins and extreme base . . . . *E. linza*
　　Frond not compressed, cavity free throughout . . . . *E. prolifera*

6   Thallus repeatedly branched with alternately long and short spine-like
    branches .... *E. crinita*
    Frond repeatedly branched without spine-like branches .... *E. clathrata*

Enteromorpha clathrata (Roth) Greville (1830), p. 181.   Fig. 21A

Type missing: according to Bliding (1963), there is no holotype of Roth's *Conferva clathrata* in Herb. Ag. nor in the Bot. Inst. of Vienna (Dr K. Warniel in letter), where the holotypes of *Conferva crinita* and *Conferva flexuosa* Wulfen ex Roth are preserved. But diagnoses and illustrations of *clathrata* by authors from the beginning and middle of the 19th century indicate that *clathrata* Roth was conceived as a repeatedly ramified (not particularly proliferous) alga with large cells arranged in rows. Island of Fimbria, Baltic Sea.

Conferva clathrata Roth (1806), p. 175.

Plants soft, light apple green to dark green, forming tufts up to 10–15 (–40) cm long; diameter of thallus increasing from base towards apex, usually repeatedly branched; cells at base and in mid-thallus unordered or arranged in longitudinal and sometimes also in transverse rows; cells in ultimate branches multiseriate or uniseriate; cells rounded, quadrate, rectangular or polygonal, (14–) 25–45 (–50) μm long; chloroplast parietal, plate-shaped in surface view with entire, lobed or dentate margin, in side view narrow with curved ends, changing position with direction of incident light; majority of cells at thallus base with more than 5 pyrenoids, usually 5–10 or more.

Reproduction by quadriflagellate zoospores, 11 × 5.5 μm and by biflagellate gametes; macrogametes, 6.9–7.8 × 3.5 μm, microgametes 5.2–6.5 × 1.8–3.1 μm.

Chromosome no.: not known.

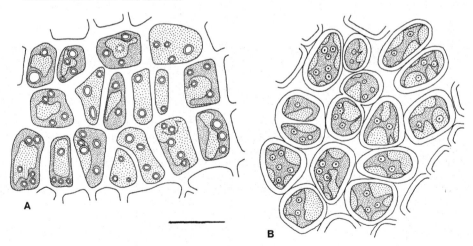

Fig. 21   *Enteromorpha* spp.
          A. *E. clathrata*. Surface view of thallus showing regular arrangement of cells. (C. Pybus). B. *E. crinita*. Surface view of thallus showing irregular arrangement of cells. (Bliding). Bar= 20 μm.

Found growing on both sheltered and exposed coasts, on rocks, stones and shells, but usually epiphytic and usually in rock pools in the littoral region, occasionally in the sublittoral; present as isolated individuals and sometimes massed as patches or more extensive carpets; found also in estuaries and brackish habitats.

Found around the whole coast of the British Isles in suitable habitats.

On Atlantic European coasts from Spitzbergen and the Polar Seas south to the Mediterranean, Baltic Sea.

On the American Atlantic coast from the Canadian Arctic south to the Caribbean and Brazil; on the American Pacific coast from Alaska south to Mexico; Iceland, Faroes; Nigeria; Australia, Tasmania.

Plants can be found throughout the year, but are most conspicuous during the summer when maximum growth occurs. Reproduction can occur at any time of the year, but mainly during the summer months. The life history is an isomorphic alternation of gametophyte and sporophyte generations; both macro- and microgametes can develop parthenogenetically.

There is considerable variation in the size and form of the plants, with the main axis sometimes conspicuous, at other times obscured by the branching. The branching may be to different degrees and the form of the branches varies; they sometimes end in a single series of cells. The different morphological forms appear to be related to environmental conditions; transitional forms and cross fertilisation occur. The cell structure is fairly constant, but may give the impression of varying because the chloroplast changes its position in the cell in relation to the direction of incident light (Bliding, 1963; Silva, 1969, 1978).

Enteromorpha crinita (Roth) J. Agardh (1883), p. 145.   Fig. 21B

Lectotype: WU (No. 70 in Herb. Wulfen. The specimen is annotated in Roth's hand '*Conf. crinita* Roth Catalecta bot. Fasc. 1 p. 162 Tab. 1 Fig. 3', see Bliding, 1963). Nr Eckwarden, East Friesland, Netherlands (see Scagel, 1966).

Conferva crinita Roth (1797), p. 162.
Enteromorpha ramulosa (Smith) Hooker (1883), p. 315.

Plants light green, up to 40 cm long, repeatedly branched with alternating long branches and short spine-like branchlets; cells at base and mid-thallus unordered or arranged in longitudinal and sometimes also transverse rows; cells rounded to quadrangular or rectangular, in lower thallus 40×30 μm, in upper part about 20×15 μm; chloroplast moves position with incident light, in surface view plate-like with entire or lobed or dentate margin, in side view rod-shaped with curved ends; majority of cells with more than 5 pyrenoids at base of plant, usually 8 or more.

Reproduction by quadriflagellate zoospres, $10–13 \times 5–8$ μm, and by biflagellate anisogametes, microgametes $6–7 \times 2–3$ μm, macrogametes $6–9.5 \times 3–4$ μm.

Chromosome no.: n = c.12 Sarma (1958). Kapraun (1970) made a chromosome count of 20 for an asexually reproducing plant of *E. ramulosa (E. crinita)* from Texas which he assumed was diploid.

Found growing on stones and on other algae, often forming mats, from the midlittoral to the sublittoral region on fairly sheltered shores; also in rock pools and on rock surfaces on gently sloping shores.

Generally distributed round the whole coast of the British Isles.

Scandinavia south to the Mediterranean and the Adriatic Sea.

On the Atlantic coast of North America from Eastern Canada to the Caribbean; on the Pacific coast of North America from Alaska south to California; Galapagos Islands; Kamchatka.

Most of the records of *Enteromorpha crinita* refer to the species as present during the late summer and winter months through to April. It may, in fact, be present throughout the year, but be confused with *E. clathrata* during the summer. The few records that are available refer to reproduction during the summer: much more information is required to describe the seasonal reproductive behaviour. The life history is an isomorphic alternation of gametophyte and sporophyte generations.

The general structure of *E. crinita* approaches closely to that of *E. clathrata*, but the presence of alternating long and short branches, the short branches spine-like, distinguishes it from the latter species. The only difficulty is that in summer the short branches may not be noticeably spine-like and then the species can be confused with *E. clathrata* (Bliding, 1963), though usually some spine-like branches are present. The most characteristic appearance occurs in the winter, though then the plants are less plentiful and may be stunted in growth. Klugh (1922) described proliferous forms of *E. crinita* which may be directly related to fluctuating salinities (see Reed & Russell, 1978).

This species has long been referred to as *Enteromorpha ramulosa* (J. E. Smith) Hooker, but there seems no doubt that *E. crinita* (Roth) J. Agardh is the correct name (see also Bliding, 1963). ▸

Enteromorpha flexuosa (Wulfen ex Roth) J. Agardh (1883), p. 152.   Fig. 22A

Holytype: WU (No. 23 in Wulfen's Herbarium as *Ulva flexuosa*, see Bliding, 1963). Duino, near Trieste, Adriatic coast.

Conferva flexuosa Wulfen ex Roth (1800), p. 188.
Ulva flexuosa Wulfen (1803), p. 1.
Enteromorpha intermedia Bliding (1955), p. 262.

Thalli characteristically slender, branched, sometimes broad, occasionally saccate, up to 60 cm or more long; branches opposite or alternate, often proliferous; cells at base of thallus in longitudinal, sometimes also in transverse rows, cells in middle region in both longitudinal and transverse rows, ultimate branches sometimes uniseriate; cells quadrangular or rectangular, 15–50 × 9–25 μm; chloroplast parietal, plate-like, sometimes forming open-ended cylinder with dentate edges; pyrenoids (1–) 2–4 (–7), few cells with single pyrenoid.

Reproduction by biflagellate gametes and quadriflagellate zoospores or by biflagellate asexual swarmers.

Chromosome no. : n = c. 16 Sarma (1958) for *E. intermedia* (*E. flexuosa*) (see Bliding, 1963). Kapraun (1970) made a count of 10 chromosomes for an asexually reproducing *E. flexuosa* from Texas.

Plants occur on both sheltered and exposed shores throughout the littoral region and in the sublittoral, also in harbours and estuaries and on salt marshes, growing on stones and on other algae or in loose masses, occasionally in freshwater.

Found round the whole coast of the British Isles in suitable habitats.
Found in Scandinavia south to the Mediterranean and the Adriatic Sea.
On the American Pacific coast from British Columbia to Central America, Galapagos Islands; Bermuda; on the Atlantic coast of North America from the Canadian Arctic south to the Caribbean; Nigeria, Ghana.

The plants are present all the year round, but in lagoons are more evident during the summer months. There is no information on the reproductive period. The life history is an isomorphic alternation of gametophyte and sporophyte generations; sometimes a succession of asexual generations occurs.

Enteromorpha flexuosa is a very variable species, it may be a simple, or only slightly branched plant or one which is highly branched and proliferous, sometimes very fine with cells in the ultimate branches uniseriate, sometimes with a main axis up to 2 cm broad, with few coarse branches. The variants may also differ in cell dimensions and in pyrenoid number, but intermediates occur. Some of the forms are difficult to distinguish from *E. clathrata*. The species *E. pilifera* Kützing (1856), *E. paradoxa* (Dillwyn) Kützing (1845) and *E. linziformis* Bliding (1960) have been given subspecific rank within *E. flexuosa* by Bliding (1963). *E. intermedia* Bliding (1955), a species he described as distinct, he later (1963) decided was identical with *E. flexuosa*.

Enteromorpha intestinalis (Linnaeus) Link (1820), p. 5.   Fig. 22B; Pl. 3

Type missing: (see Bliding, 1963). Swedish Lapland.

Ulva intestinalis Linnaeus (1753), p. 1163.
Enteromorpha compressa (Linnaeus) Greville (1830), p. 180.

Thallus light to dark green, completely tubular, increasing in width from base to mid-thallus, often more or less stipitate, 10–30 cm or more long, up to 3 cm or more broad, unbranched or branched; cells irregularly arranged at base to mid-thallus, small areas of at least 10–12 cells regularly arranged in mid-thallus not uncommon, rarely in circles round a centre of growth; cells (12–) 20–25 × 11–17 μm, polygonal, rectangular or rounded; chloroplast usually hood-shaped in surface view and located at the upper end of the cell, shape modified by injury and by cell division, assumes lateral position at night; majority of cells with 1 pyrenoid, few with 2.
Reproduction by quadriflagellate zoospores and anisogamous biflagellate gametes; gametes sometimes parthenogenetic.

Chromosome no.:   n = 10   McArthur & Moss (1978).
            n = 9/10   Sarma (1958).
            n = 10,   2n = 20 Ramanathan (1939) for *E. compressa* var.
                    lingulata.

Found on both sheltered and exposed coasts, on rock, in pools, on stones, boulders,

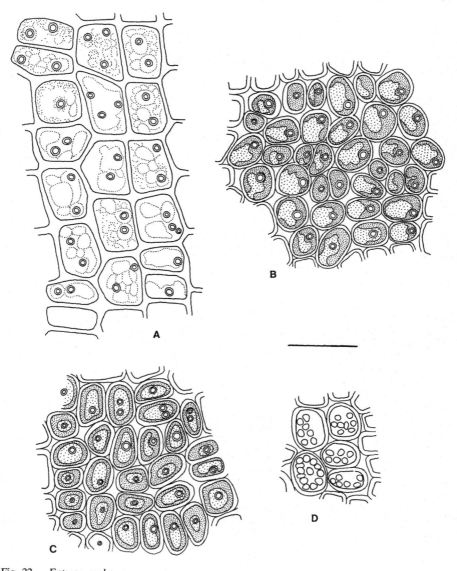

Fig. 22   *Enteromorpha* spp.
A. *E. flexuosa*. Surface view of mid thallus region showing regular arrangement of cells. (Ferry Bridge, Dorset; Oct. 1980). B. *E. intestinalis*. Surface view of thallus showing irregular arrangement of cells and polar position of chloroplast. C, D. *E. prolifera*. Surface view of vegetative (C) and reproductive cells (D). (Bliding). Bar= 20 μm.

shells, coastal installations and on other algae from the upper littoral to the sublittoral; subspecies *intestinalis* is more common towards upper shore positions, subspecies *compressa* towards lower shore positions; also found in estuaries and lagoons under brackish water conditions, in freshwater lakes and canals under alkaline conditions, sometimes free-floating.

Common all round the coasts of the British Isles.
More or less world-wide in its distribution.

Plants can be found in a reproductive condition at all times of the year; maximum development and reproduction occur during the summer months especially towards the northern end of the distribution of the species. The life history is an isomorphic alternation between gametophytic and sporophytic generations, but can be modified by environmental conditions (Burrows, 1959; Moss & Marsland, 1976; Reed & Russell, 1978).

For comments on form variation see discussion under genus *Enteromorpha* p. 82.

The only character separating *E. intestinalis* and *E. compressa,* as originally described and interpreted since then, is that the former species is unbranched and the latter branched. Bliding (1963) reported, however, that when he applied an inter-sterility barrier criterion to the separation of the two species, he had to include a few branched plants in *E. intestinalis* and a few unbranched plants in *E. compressa.* Treating the branched and unbranched members of this complex together Silva (1969) and Silva & Burrows (1973) found that branching was expressed mainly in plants at the lower littoral levels, becoming less common in the upper part of the shore. Silva found that, in laboratory culture experiments, unbranched plants could give rise to branched offspring and branched plants to unbranched offspring. This evidence supports the suggestion that branching is a character which is polygenetically controlled. Further it was shown that there was a partial interfertility between branched and unbranched plants, plants from the lower part of the shore crossing less readily with those from the upper part than with plants from the nearer levels. Bearing all of this information in mind there seems no basis, either morphological or genetic, for distinguishing two species within this complex. The data are in accord with the presence of two ecotypes: These can be treated taxonomically as two subspecies:

> *Enteromorpha intestinalis* (L.) Link subspecies *intestinalis* – unbranched
> *Enteromorpha intestinalis* (L.) Link subspecies *compressa* – branched

Distinguishing the subspecies presents some of the problems of distinguishing the original species as there is the practical problem in deciding whether or not a plant is branched when the expression of branching is slight. Mature plants collected from the field can be found unbranched or with branches in any of the developmental stages from a single cell or short uniseriate filament to a branch reproducing the character of the main axis. The aggregate species *Enteromorpha intestinalis* (L.) Link would seem to be the most useful to apply until much more work has been carried out on this complex from a genetical point of view.

Enteromorpha linza (Linnaeus) J. Agardh (1883), p. 134.   Pl. 4

Type missing: 'in Oceano', probably England, see Scagel, 1966, p. 54.
Ulva linza Linnaeus (1753), p. 1163.

Thalli light or dark green, unbranched, up to 30 cm or more long, up to 5 cm or more wide, often with frilled margin, tapering gradually or abruptly into distinct stipe below, width of thallus much greater in middle than base of thallus; walls of central frond cavity closely adpressed, free only at margins and extreme base; cells quadrangular, about 15 μm diameter, or rectangular 14 × 22 μm, cells at base of thallus with rhizoidal elongations, arranged in longitudinal rows, sometimes also in transverse rows; cells in mid-thallus arranged predominantly in longitudinal and transverse rows; chloroplast parietal, appearing ring-shaped in surface view at least in basal region; pyrenoid usually single in basal cells; forms with up to 3–5 pyrenoids in some cells recorded.

Reproduction by quadriflagellate or biflagellate zoospores; also by biflagellate isogametes.

Chromosome no.: 2n (?) = c. 24–25 Levan & Levring (1942).
                 n (?)   =   10 Comps (1960).
                 n   =   12 Niizeki (1957).

Found in rock pools and on rock surfaces, forming isolated patches or more extensive carpets throughout the littoral region and in the sublittoral; frequent on gradually sloping shores, but not confined to them; occasionally found under brackish conditions, but usually marine.

Found around the whole coast of the British Isles; not uncommon.

Russian Arctic, Scandinavia south to the Mediterranean.

North American Pacific coast from Alaska to Mexico, Chile; North American Atlantic coast from the Maritime Provinces of Canada south to the Caribbean, Bermuda; Iceland, Faroes; southern Australia.

Probably present throughout the year, but most conspicuous by its bright green growth during the spring months. It is known to reproduce during the summer, but information on the reproductive period is incomplete.

The life history of *Enteromorpha linza* may possibly be altered by environmental conditions. This has been suggested by Arasaki & Shihiva (1959). In Europe it appears that it involves only an asexual cycle. The plants, however, may have both quadriflagellate and biflagellate zoospores. Levan & Levring (1942) found quadriflagellate zoospores with a chromosome number of about 24 and they assumed the plants were diploid. Comps (1960) found plants on the coast of France which had quadriflagellate zoospores with a chromosome number of 10, this number being passed on to the next generation without meiosis occurring. Niizeki (1957) working in Japan found biflagellate zoospores with a chromosome number of 12 and again without meiosis between generations. An alternation of gametophyte and sporophyte generations, as well as an asexual cycle, was reported in Japan by Arasaki & Shihiva (1959) though Bliding (1963) expressed doubts as to whether the plants belonged to *E. linza* as it is known in Europe. An alternation of isomorphic generations was reported for this species by Scagel (1966) in British Columbia and northern Washington. The species needs further investigation.

Enteromorpha prolifera (O.F. Müller) J. Agardh (1883), p. 129.   Fig. 22C, D

Type: C? Germany, near Nebel, North Friesland.

Ulva prolifera O.F. Müller (1878), p. 7, pl. 763.
Enteromorpha ahlneriana Bliding (1933), p. 248.
Enteromorpha torta (Mertens) Reinbold (1893), p. 205.

Thalli forming a tangled mass of threads, hollow, unbranched or sparsely branched, and more or less the same diameter throughout, diameter of cavity about 12 μm; or thalli distinct, increasing in width from base towards middle region, from a few cm to more than 2 m long, 1–2–30 or more cm broad; thalli usually branched, often proliferous; branches sometimes tapering at base and apex; thalli sometimes with trabeculae extending into central cavity; cells at base in longitudinal rows, in middle region in longitudinal and also transverse rows, distal cells often disordered; cells in narrow thalli in 3–12 distinct longitudinal rows, quadrangular or rectangular, 12–28 × 11–16 μm; cells in separate broad thalli 9–19 × 7.5–15 μm; chloroplast a parietal plate overlapping more than one wall of cell and appearing ring-shaped in surface view with one large central pyrenoid, a few cells with 2–3 pyrenoids.

Reproduction by quadriflagellate zoospores 11.0 × 5.2 μm, and by anisogamous biflagellate gametes, microgametes 5.9 × 2.4 μm, macrogametes 7.9 × 4.4 μm.

Chromosome no.: not known.

Most common on gently sloping shores on sheltered coasts; plants epilithic or epiphytic throughout the littoral region, extending into muddy estuaries and brackish water; often forming tangled carpets on muddy substrates; reported growing on rocks, shells and other solid objects in rather sheltered situations at the lowest littoral level; also found free-floating.

Common all round the coast of the British Isles.
Scandinavia south to the Atlantic coast of Spain and Portugal, Baltic Sea.
Pacific coast of America from Alaska to Mexico and Peru; Atlantic coast of America from the Canadian Arctic south to the Caribbean and Brazil; southern Australia, Sri Lanka, Japan.

Plants can be found at all times of the year, but in the more northerly parts of its range it occurs only during the summer months. Swarmer production has been reported on the south coast of Britain from September to January. The life history is an isomorphic alternation between gametophyte and sporophyte generations.

Enteromorpha prolifera grows in a wide range of habitats and is a very variable species. Since *E. torta* has been recognised as belonging to it, it ranges in morphology from long narrow threads with a wall of 3–12 rows of cells of equal diameter throughout to thalli of the more normal *Enteromorpha* shape, becoming broader from the base upwards and reaching a length of as much as 2 m and a breadth of up to 30 cm under the influence of freshwater at the mouths of small streams. The cell arrangement and form of the chloroplast, however, remain fairly constant.

Enteromorpha ahlneriana Bliding falls within the *E. prolifera* complex. It was separated by Bliding (1933) on small variations in cell size and arrangement, size of zoospores and the life history which apparently involves only an asexual cycle. Under certain conditions, often in brackish water and on mud flats in estuaries, *E. prolifera* occurs in a very narrow form, very sparsely branched or even unbranched. In this form it approaches very closely *E. torta*. According to Bliding (1963) this form can be distinguished from *E. torta* by the narrow central cavity of the latter species. The two species have the same cell arrangement and cell structure. Polderman (1975) in his work on two brackish polders in the Netherlands, found the diameter of the central cavity to be variable and he decided that there were no reliable morphological characters to distinguish *E. prolifera* from *E. torta*. *E. torta* is stated by Bliding (1963) to be reproductively isolated from *E. prolifera*, but because they merge morphologically and distinction is extremely difficult, Polderman (loc. cit.) has been followed in including *E. torta* in *E. prolifera*.

Enteromorpha ralfsii Harvey (1851), p. 339.

Type: according to Bliding (1963), the specimen is not in Harvey's herbarium at TCD and J. Agardh's material of this is mostly *Enteromorpha torta*.

Thallus of tangled threads many cm long, diameter the same throughout, only occasionally branched; branches short; thalli of 3–8 distinct longitudinal rows of cells surrounding central cavity about 8 μm diameter; cells quadrangular about 15 μm diameter or rectangular 23 (–32) × 17 μm, height in cross section about 19 μm; majority of cells with more than 1 pyrenoid, usually 2–8.
Reproduction by quadriflagellate zoospores.

Chromosome no.: not known.

Found on muddy shores forming tangled masses over the substrate, or as isolated threads mixed with *E. prolifera* and other algae in the littoral region; also in harbours and estuaries.

There are few records of this alga for the British Isles, but it does not appear to be confined to a particular region. It may have been overlooked.
Devon, Caernarvon, Co. Cork, Shetland, Northumberland, Isle of Man, Isle of Wight.
Atlantic coast of France.

The plants are reported to occur and to reproduce mainly in the summer. The life history involves only an asexual cycle. Bliding (1963) reported that the zoospores from field material do not germinate normally, but two or more fuse together. The germination percentage was very low. This raises the question of the status of this species and it needs further investigation.

Harvey's (1851) illustration of *E. ralfsii*, while indicating the large size of the cells, shows clearly a single pyrenoid in each cell and he also commented on this. In the plant, as it is now known, this condition is rare. Harvey received his material from Mr Ralfs under the name *Enteromorpha percursa*.

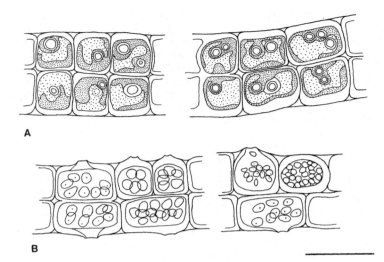

Fig. 23   *Percursaria percursa*
Portions of vegetative (A) and reproductive filaments (B). (Ferry Bridge, Dorset; Oct. 1980; Stornoway, Outer Hebrides; Nov. 1971). Bar= 20 μm.

PERCURSARIA Bory

PERCURSARIA Bory (1823), p. 393.

Type species: *Percursaria percursa* (C. Agardh) Rosenvinge (1893), p. 963.

Zignoa Trevisan (1842), p. 50.
Tetranema Areschoug (1850), p. 418, non *Tetranema* Bentham (1843) nom. cons.
Diplonema Kjellman (1883), p. 302, nom. illeg., non De Notaris (1846).

Thallus of unbranched filaments, biseriate with cells usually lying opposite or occasionally uniseriate, arising from margin of one-layered disc or directly without formation of a disc; chloroplast a parietal plate or irregular ring often broader on internal walls, with 2–3 pyrenoids.
    Reproduction by quadriflagellate zoospores and biflagellate gametes.
    Percursaria is a distinct and readily recognised genus.

Only one species in the British Isles.

Percursaria percursa (C. Agardh) Rosenvinge (1893), p. 963.   Fig. 23

Lectotype: LD (Herb. Agardh: Bliding (1963) reported '*Conferva percursa* C. Agardh, accompanied by a drawing in C. Agardh's hand giving a characteristic image of one *Percursaria* thread of two cell series with the annotation "filis simplicibus Conferva percursa mihi",' as No. 13617 while Dixon, who examined material at Lund, gave No. 13716 as did Scagel, (1966). Hofmansgave, Denmark.

Conferva percursa C. Agardh (1817), p. 87.

Plants of unbranched filaments, biseriate with cells usually lying opposite or occasionally uniseriate; filaments arising from margins of one-layered disc or directly without disc; cells up to 25 × 18 μm, 1–3 × long as broad; chloroplast a narrow parietal ring sometimes broadening on inner wall of cell, or a broad plate in dim light; pyrenoids 2–3; filaments pale green.

Reproduction by quadriflagellate zoospores or biflagellate gametes; zoospores 7.0–8.0 × 4.5–5.0 μm: microgametes 5.5–7.0 × 2.0 μm, macrogametes 5.0–7.0 × 3.0–4.0 μm; plants dioecious or monoecious.

Chromosome no.: not known.

Found growing in tangled masses or intertwined with other algae on mud in estuaries, on rocks, or in somewhat brackish pools; also on salt marshes and may form floating masses in closed littoral lagoons in more or less brackish water; may become luxuriant in well sheltered shallow bays.

The species occurs all round the coast of the British Isles in suitable habitats.

Spitzbergen and Scandinavia south to the Atlantic coast of France, Baltic Sea, Novaya-Zemlya.

Eastern Canada south to New Jersey on the Atlantic coast of North America; Alaska south to California on the Pacific coast of North America; Japan.

The species is recorded as present and fertile at all times of the year in the British Isles, but the type of reproductive body is not always recorded. Recently plants have been found in October producing gametes copiously and of two sizes from the same filaments in the Fleet, Dorset, after the area has been flooded with low salinity water. Kornmann (1956) reported finding only zoosporangia in field material. The life history is an isomorphic alternation of gametophyte and sporophyte generations, but biflagellate swarmers from individual plants may develop to produce filaments which form either bi- or quadriflagellate swarmers (Kornmann, 1956).

ULVA Linnaeus

ULVA Linnaeus (1753), p. 1163.

Type species: *Ulva lactuca* Linnaeus (1753), p. 1163.

Ramularia Rousell (1806), p. 98.
Phycoseris Kützing (1843), p. 296.
Letterstedtia Areschoug (1850), p. 1.
Lobata Chapman (1952), p. 48.

Thallus a flat, entire or divided lamina with flat, fluted or toothed margin, consisting of two closely coherent layers of cells dividing only in transverse and longitudinal radial planes; lamina attached by holdfast formed from rhizoids growing down from lower cells of thallus, with or without distinct stipe; germling passing through tubular 'Enteromorpha' stage. Reproduction by quadriflagellate and sometimes also by biflagellate zoospores and by biflagellate gametes, isogamous; gametangia formed from any cell of thallus except basal cells.

Three species of *Ulva* have been recorded for the British Isles. Eight species were recorded by Bliding (1968) for Europe and 5 species including a new species by Koeman & van den Hoek (1981) for the Netherlands. It is possible that more than three species occur in Britain, but so far they have not been distinguished.

Papenfuss (1960) discussed in some detail the nomenclature of *Ulva* L. (1753), typified by *Ulva lactuca* L. and suggested that Linnaeus probably drew his diagnosis of *Ulva* from his species *U. intestinalis*. He said 'On this basis *Ulva* becomes a synonym of the conserved name *Enteromorpha* Link 1820 which also has *U. intestinalis* as lectotype (cf. Silva, 1952, p. 294)'. Papenfuss proposed that *Ulva* should be conserved in the sense of Thuret (1854, p. 28), with *U. rigida* (C. Agardh) Thuret as lectotype species. So far this proposal has not been accepted.

In describing the thalli of the species of *Ulva* the divisions of the blade used by Koeman & van den Hoek (1981) have been followed. These are:

basal region – the zone between stipe and upper limit of the rhizoidal cells occurring interspersed among normal vegetative cells.

middle region – the zone situated between about $\frac{1}{3}$ and $\frac{1}{2}$ of the thallus length from the stipe.

apical region – the zone situated between about $\frac{2}{3}$ of the thallus length from the stipe and the thallus apex.

KEY TO SPECIES

1　Majority of cells with one pyrenoid; dark coloured rhizoidal cells in upper basal region approximately the same size as vegetative cells in surface view. . . . . 　2

　　Majority of cells with two or more pyrenoids; dark coloured rhizoidal cells in upper basal region larger than normal cells; margin of frond, especially in basal region often with microscopic teeth, ... *U. rigida*

2　Fronds light to dark green; cells in slightly curved rows; mean cell dimensions $18 \times 13.5 \ \mu$m in surface view . . . . *U. lactuca*

　　Fronds olive green; cells irregularly arranged: mean cell dimensions $24 \times 18 \ \mu$m in surface view . . . . *U. olivascens*

Ulva lactuca Linnaeus (1753), p. 1163.　　　　　　　　　　　　　Fig. 24; Pl. 5

Holotype: LINN (No. 1275–24 Specimen marked '5'. 'The specimen agrees well with Linnaeus's description of the species, and furthermore '5' is the number assigned to *U. lactuca* by Linnaeus in the *Species Plantarum*': Papenfuss, 1960).

Thallus soft except at base where texture is firmer, light to dark green, a few cms to 1m or more long, more or less rounded, usually divided more or less deeply, with fluted margin, 50–90 (–120) μm thick, attached by disc-like holdfast, sometimes loose lying; basal region of frond with longitudinal ridges formed by bundles of rhizoids which are internal extensions of basal cells; cells at base of frond rounded, arranged in more or less curved rows, interspersed with dark coloured rhizoidal cells larger than vegetative cells, in upper basal region rhizoidal cells approximately the same size as vegetative cells in surface view; cells in mid-frond quadrangular, rectangular or irregularly polygonal with rounded corners, (11–) 15 (–27) × (8–) 10 (–20) μm; chloroplast a

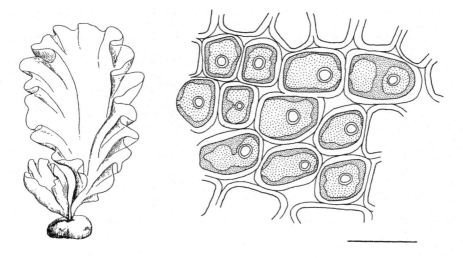

Fig. 24   *Ulva lactuca*
A. Habit of plant (after Newton, 1931). Bar= 30 mm. B. Surface view of vegetative cells. Bar= 20 μm.

lobed parietal plate with 1, rarely 2–3 pyrenoids, rhizoidal cells with several pyrenoids.

Reproduction by quadriflagellate zoospores and by biflagellate anisogamous gametes; plants dioecious; zoospores 10–12 × 5.5–6.0 μm, microgametes 7–9 × 2.5–4 μm, macrogametes 8–11 × 4–5 μm; vegetative reproduction by small fragments of frond acting as centres of growth for new fronds.

Chromosome no.:   n = 10 Carter (1926).
                  n = 13, 2n = 26 Föyn (1934).
                  n = 10, 2n = 20 Sarma (1958).

On rocks, in pools and on other algae from the upper littoral to the sublittoral, in estuaries and on salt marshes on small stones and shells, and loose lying over sand and mud.

Common all round the coast of the British Isles.
More or less world-wide in its general distribution.

Plants can be found at all times of the year with maximum development in summer. Reproduction occurs at all times of the year, but especially during the summer months. Vegetative reproduction common in loose lying populations. The life history is an alternation of isomorphic gametophyte and sporophyte generations.

The fronds vary very much in size from a few centimetres to flat expanses of up to one metre or more in length and breadth and the colour varies from light to dark green. Plants from the sublittoral growing round the mouth of a drain in Port Erin, Isle of Man, were very large and dark green. Plants also reach a very large size growing unattached in polluted harbours and estuaries; the colour of these is much lighter green and often quite pale in exposed parts of the shore. The growth of *Ulva lactuca* is encouraged by the presence of ammonia in the water and it may form extensive sheets, e.g. in Poole Harbour and also on the open coast if the plants occur below rotting piles of drift weed. Large fronds lying on mud tend to be fenestrate due to predation by animals. Frond shape varies from strap-shaped with wavy margins and attenuated base to more or less circular, undivided or divided into deep or shallow segments. The texture is soft or sometimes firm; in the latter case the microscopic teeth sometimes present on the frond margins of *Ulva rigida*, are always lacking.

When, at Papenfuss's request, the holotype of *Ulva lactuca* in the Linnean Herbarium was soaked out, it was found (Mrs Butler in a letter to Papenfuss) that the base of the frond 'seemed thick and fleshy, the edge of the lower part of the frond was dentate with quite large teeth'. This suggests that the specimen, which in sectioning was shown to be distromatic, was not *U. lactuca,* but *U. rigida* as these species are now understood. Bliding (1968) was, however, of the opinion that Papenfuss's description of the type material fits exactly the characters of what is now known as *Ulva lactuca.*

Ulva olivascens Dangeard (1961), p. 22.   Fig. 25

Type: BORD. Roscoff, Finistère, France.

Ulva olivacea Dangeard (1951), p. 27. nom. illeg., non *U. olivacea* Hornemann (1810).

Thallus flat, firm, distinctly olive green in colour, up to about 30 cm long; lamina with small rounded perforations and fluted margin, sometimes markedly laciniate, 40–60 μm thick in middle region, up to 100 μm or more at base; basal region with longitudinal ribs formed from bundles of descending rhizoids; in upper basal region rhizoidal cells approximately the same size as vegetative cells in surface view; cells isodiametric, polygonal or rectangular with rounded corners, 24–35 × 18–24 μm in middle region, more or less rounded and rather smaller in apical region; chloroplast a parietal plate, only slightly lobed, with minute perforations, lying along outer wall of cell or on upper part of side wall; pyrenoid single, only occasionally 2–3.

Reproduction by biflagellate or quadriflagellate asexual swarmers, formed in slightly enlarged vegetative cells; marginal layer of cells may remain sterile; Gayral (1962) reported anisogamous gametes and zoospores for plants from the coast of Morocco.

Chromosome no.: not known.

Found on rocks and harbour installations in places shaded from direct sunlight on the lower littoral and in the sublittoral.

Fig. 25   *Ulva olivascens*
A. Habit of plant. (Cornwall; 1987). Bar= 20 mm. B, C. Surface views of vegetative thalli showing regular (B) and irregular (C) arrangement of cells. Figures drawn from Bliding's Plymouth slide, Wembury, Devon; Oct. 1960). Bar= 20 μm. D. Surface views of reproductive cells with enclosed swarmers. Bar= 20 μm.

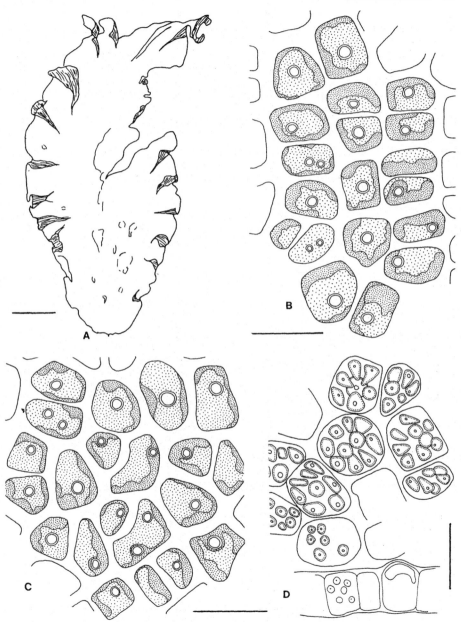

So far found only rarely in the British Isles, but may be locally abundant; Channel Islands, Devon, Cornwall, Co. Clare.

Atlantic coasts of France and Spain, Mediterranean.

Young plants may appear as early as January, more common in spring, summer and early autumn, but full seasonal occurrence is uncertain. The reproductive period is not known for the British Isles, but quadriflagellate zoospores were reported for plants at Roscoff, France, in August and September (Dangeard, 1961). The life history is incompletely known. The plants appear to reproduce only by bi- or quadriflagellate zoospores, but the fact that anisogametes have been found in Morocco (Gayral, 1962) may mean that the life history varies with environmental conditions. This needs further investigation.

Apart from the fact that the fronds may vary in shape from more or less orbicular to elongate and reniform, little is known of variation for this species.

Ulva rigida C. Agardh (1822), p. 410.   Fig. 26

Lectotype: LD (designated by Papenfuss, 1960, p. 305; specimen No. 14294 in the Agardh Herbarium, sent to Agardh by Cabrera, probably Cadiz, Spain.

Thallus a dark green flat lamina, usually 10–15 cm long, but up to one metre, with

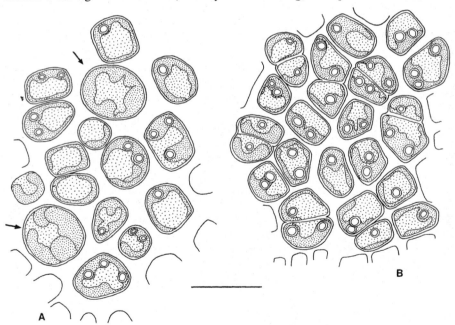

Fig. 26   *Ulva rigida*
A. Surface view of basal thallus region. Note larger rhizoidal cells (arrowed). B. Surface view of mid thallus region. (Cornwall). Bar= 20 μm.

fluted margins and often deeply divided, stiff in texture, narrowing abruptly to distinct stipe, attached by rounded disc; margin of frond often with microscopic tooth-like projections, especially at base; base of lamina with longitudinal ribs formed by bundles of rhizoids; cells in upper basal region with unordered, rounded cells interspersed between dark-coloured rhizoidal cells up to twice the diameter of vegetative cells in surface view; cells in middle to apical region arranged in longitudinal and also transverse rows, polygonal with rounded corners, (11–) 15 (–29) × (7–) 9 (–14) μm; marginal cells in apical region spherical; lamina 40–120 μm thick; chloroplast parietal with 2–4 pyrenoids, often situated at one side of cell so as to appear somewhat hood-shaped in surface view.

Reproduction by quadriflagellate zoospores and by anisogamous biflagellate gametes; plants dioecious.

Grows attached to rock surface, stones, woodwork and other algae or unattached in harbours and sheltered bays and lagoons; also on exposed shores, throughout the littoral zone and into the sublittoral, extending into brackish water; it occurs mainly in deep pools on the lower part of the littoral and in the sublittoral and sublittoral fringe.

Generally distributed round the whole coast of the British Isles, probably more common than indicated by the records.

Common on the Mediterranean coast and extending from the Canary Islands to the English Channel, less common north to Scandinavia.

Arctic Canada, Maritime Provinces of Canada, American Pacific coast from Alaska to southern California, Chile; South Africa; Sri Lanka, southern Australia, New Zealand.

Plants are common during the summer months and may be present all the year round. Reproductive plants have been found in the summer but the seasonal behaviour in relation to reproduction is not known. The life history is an alternation of isomorphic gametophyte and sporophyte generations, but both macro- and micro-gametes can develop parthenogenetically.

There is a size variation in the lamina from a few centimetres broad in extremely exposed sites to fronds one metre or more broad in sheltered bays or harbours and in the sublittoral. In some of the brackish lagoons in southern France the plants are found lying loose; these are highly perforated and reproduce vegetatively (Bliding, 1968).

There is considerable variation in size measurements, given by different authors, for the zoospores and gametes.

	Bliding 1968	*Koeman & van den Hoek 1981*
Zoospores	9.1–10.6 × 4.6–6.0	(13–) 13.5 (–14) × 7–7.5 (–8) μm
Micro-gametes	5.3– 6.3 × 2.7–3.1	(7–) 8 (– 9) × 3–4 (–5.5) μm
Macro-gametes	6.2– 7.4 × 3.1–4.5	(10–) 11 (–12) × (3.5–) 4 (–6) μm

The cause of this variation is not clear.

CHAETOPHORACEAE De Toni & Levi-Morenos

CHAETOPHORACEAE De Toni & Levi-Morenos (1888), p. 131.

Plants freshwater, brackish and marine; microscopic branched filaments epi-endophytic, epi-endozoic, prostrate only or with limited erect system, free or forming pseudoparenchyma, at times more or less enclosed in a gelatinous envelope; cells uninucleate, with or without setae; chloroplast parietal, more or less perforate, sometimes reticulate with one to several pyrenoids; pyrenoids sometimes absent. Reproduction by tri- or quadriflagellate zoospores and by isogamous or anisogamous biflagellate gametes; plants monoecious or dioecious; vegetative reproduction by formation of akinetes or aplanospores and by fragmentation.

Life history: an isomorphic alternation of gametophyte and sporophyte generations or a heteromorphic alternation involving a unicellular sporophyte.

The marine members of the Chaetophoraceae are regarded as a very coherent group of plants by O'Kelly (1980) on the basis of the following characters:

1 The basic unit of thallus construction is a branched filament in which apical growth predominates. The filaments may remain free or may unite into a pseudo-parenchyma. Thallus development may be prostrate or heterotrichous. Wholly erect thallus forms are absent, as is true parenchyma. Setae are present in most species and no other type of hair-like appendage is produced.

2 Vegetative cells are uninucleate and possess a single parietal chloroplast with one or more pyrenoids. The differentiation of the chloroplast inner membranes into grana and stroma lamellae is poor. The pyrenoid is traversed by 1–3 straight thylakoids. Mitochondrial cristae are flattened and elongate. The cell wall is composed of a fibrillar layer and stains readily with histochemical dyes for cellulose.

3 Sporangia and gametangia are identical to one another in structure and develop-ment (except in *Bolbocoleon piliferum, Eugomontia sacculata* and *Pseudopring-sheimia confluens*). They are cylindrical to flask-shaped and have a single apical exit papilla. The mother cell protoplast undergoes sequential cleavages to produce 4–64 swarmers. The wall of the mature sporangium or gametangium is bi-layered, the inner wall (capsule) protruding through the outer wall to form an exit papilla. The papilla ruptures explosively at the moment of swarmer release. No vesicle surrounds the swarmers.

4 The sexual life history involves an alternation of isomorphic phases (except in *Pseudopringsheimia confluens*). Meiosis takes place in the sporangia produced by the diploid phase. Zoospores are quadriflagellate and germinate to produce monoecious or dioecious gametophytes. The gametes are biflagellate. Anisogamy is typical, with isogamy present in a few cases. Gamete pairing occurs without an intervening aggregation phase and syngamy is complete in a matter of seconds. Sporophytic plants develop from the zygotes. There is no resting stage in the life history. Reproduction by quadriflagellate zoospores occurs, but biflagellate zoo-spores and parthenogenetically developing gametes do not. Chromosome numbers are low, usually falling between n = 4 and n = 12.

5 The unique chloroplast pigment siphonoxanthin is found in many marine chaetophoracean species. The other xanthophyll pigments present are identical to those known for other green algae. Secondary xanthophylls are apparently absent.

On the basis of these characters, O'Kelly considers that the marine genera of the

Chaetophoraceae form a clearly homogenous group, distinct from the freshwater forms for which he uses the name Ulvellaceae of Schmidle (1899, emend.). *Eugomontia sacculata* and *Pseudopringsheimia confluens,* with their distinct heteromorphic life history patterns, *Codiolum*-like sporangial structure, lack of setae and isogamous sexual reproduction, he excludes from the family and suggests that they are best placed within the Ulotrichales, where they may constitute a distinct family between Ulvellaceae and Monostromataceae. *Bolbocoleon piliferum* is retained within the Ulvellaceae, having few features separating it from the other members of the family. *Phaeophila dendroides* is structurally identical with members of the Ulvellaceae, but has developed free-nuclear divisions within the sporangia by simultaneous rather than sequential cleavage. There is in addition a membranous 'plug' at the apex of the sporangial neck which opens to the environment well before the release of spores (O'Kelly & Yarish, 1980); on these grounds *Phaeophila* is excluded from the Ulvellaceae. On the ground that the sporangial development of *Phaeophila dendroides* is similar to that of two chaetosiphonacean species, he suggests placing this species within the Chaetosiphonaceae.

A number of difficulties arise in making the changes suggested above and apart from placing *Eugomontia sacculata* in the Ulotrichaceae, they have not been made here, though they will have to be considered in the future.

ACROCHAETE Pringsheim

ACROCHAETE Pringsheim (1863), p. 330 (reprint p. 8).

Type species: *A. repens* Pringsheim (1863), p. 330 (reprint, p. 8).

Plants consisting of microscopic, irregularly, often richly branched filaments, epi- or endophytic, later forming loose pseudoparenchyma with free filaments at periphery, and prostrate filaments giving rise to short erect filaments; erect filaments sometimes absent; hairs, produced as extensions of upper walls of cells formed at ends of erect filaments or directly on cells of prostrate system; chloroplast parietal, disc or plate shaped, occasionally reticulate, sometimes reported to have internal projections, covering almost all of cell wall, but area variable, one to several pyrenoids, sometimes with one pyrenoid larger and more prominent than the others.

Reproduction by biflagellate or quadriflagellate swarmers formed in sporangia terminating erect filaments; swarmers released in common mucilage envelope.

Two species of *Acrochaete, A. repens* and *A. parasitica* have in the past been attributed to the genus. *A. parasitica* Oltmanns (1894) differed from *A. repens* in supposedly being parasitic on its host, species of *Fucus.* South (1968) doubted whether this supposed parasitic nature was a valid basis for establishing a second species of *Acrochaete.* Nevertheless in 1974 he kept the two species separate with no reference to their possible conspecificity. Nielsen (1979) found no real evidence of parasitism for plants answering to the original description of *A. parasitica* found in *Fucus vesiculosus* and *F. serratus.* The fact that she was able to take *A. parasitica* into laboratory culture indicated that there was no obligate parasitism. In culture Nielsen (1979) found isolates of *A. repens* and *A. parasitica* to be 'all alike'. She therefore included *A. parasitica* as a synonym of *A. repens.* This has been accepted here.

Only one species in Britain.

Acrochaete repens Pringsheim (1863), p. 330 (reprint p. 8).

Lectotype: original illustration (Pringsheim, 1863, pl. XIX; reprint pl. II) in absence of material (see South, 1974). Helgoland, Germany.

Acrochaete parasitica Oltmanns (1894), p. 208.

Plants filamentous, irregularly and richly branched, prostrate system giving rise to erect branches of 1–5, generally few cells, terminating in fine hairs or sporangia; erect branches sometimes absent; hairs with broad bases produced as extensions of upper cell walls, not from special hair-bearing cells, hairs also produced from prostrate cells; cells variable in shape, 7–12 μm diameter, cell length variable, cells becoming thread-like in deeply penetrating filaments; chloroplast a parietal plate or disc-shaped or reticulate, sometimes with internally extending prolongations; 1–3 (–6) pyrenoids, one sometimes more prominent than others.

Reproduction by biflagellate swarmers or by quadriflagellate swarmers, formed by progressive division of mother cell contents; swarmers with red eye spot; sporangia terminal on erect branches, elongate- ovoid or clavate.

Chromosome no.: n = 13 Kermarrec (1970).

Plants endophytic or epi-endophytic in the tissues of various Phaeophyceae, chiefly *Chorda filum* and *Fucus* species. Other reported hosts: *Ralfsia, Laminaria, Leathesia, Spermatochnus, Acrothrix, Stictyosiphon, Scytosiphon, Chordaria*; also reported on leaves of *Zostera* species.

Probably more common in the British Isles than is indicated by the records.
Dorset, Devon, Anglesey, Isle of Man, Cumbrae, Northumberland, Co. Wexford.
Atlantic coast of Europe from Scandinavia to France, Baltic Sea, Helgoland.
Greenland, Atlantic coast of North America from Newfoundland south to S. Massachusetts.

Plants of *Acrochaete repens* can be found at almost any time of the year and are reproductive at different periods in different localities. Swarmers reported over the winter period, from November to April in Anglesey (South, 1968), plants sterile from November to May and reproductive in November in Eastern Canada (South, 1974), plants fertile in August in Massachusetts (South, 1974); swarmers reported in plants collected in July in northern Norway (Jaasund, 1965) and in winter in the Baltic Sea (Lakowitz, 1929). The life history is not known; swarmers have been reported as giving rise directly to a filamentous phase, but biflagellate swarmers may have a gametic function (Nielsen, 1979). Two sizes of biflagellate swarmers have been reported, though the smaller colourless biflagellate swarmers are thought by South (1974) to be fungal in origin. O'Kelly (1980) found only quadriflagellate zoospores for plants in the USA and queried the existence of biflagellate swarmers. The life history of this species obviously still needs attention.

South (1968) found evidence to suggest a pseudoparasitic effect of *Acrochaete repens* in *Chorda filum*. This effect was also found for plants originally described as *Acrochaete parasitica* Oltmanns (1894). No such effect was found by Nielsen (1979) for plants growing in species of *Fucus*.

BOLBOCOLEON Pringsheim

BOLBOCOLEON Pringsheim (1863), p. 330 (reprint p. 8).

Type species: *B. piliferum* Pringsheim (1863), p. 330 (reprint p. 8).

Plants filamentous, prostrate, irregularly branched, creeping superficially on a variety of hosts; cell division principally in terminal region of filament; erect system reduced to bulb-shaped hair-bearing cells with well developed chloroplasts, seldom separated from prostrate system by further cell divisions; hairs sometimes in dense clusters, not sheathed except following injury; chloroplast parietal, plate-like or perforated with 1–10 pyrenoids; reproduction by quadriflagellate or biflagellate swarmers.

Only one species.

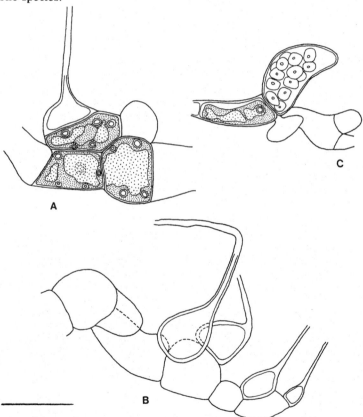

Fig. 27   *Bolbocoleon piliferum*
A-C. Portions of branched filaments showing bulbed-shaped hair-bearing cells (A, B) and sporangial cell (C). Bar= 20 μm

Bolbocoleon piliferum Pringsheim (1863), p. 330 (reprint p. 8). Fig. 27

Lectotype: original illustration (Pringsheim, 1863, pl. XVIII, Fig. 1: reprint pl. I, Fig. 1), see South (1969). Helgoland, Germany.

Thallus consisting of prostrate system of irregularly branched filaments; cell shape irregular, cells up to 15 μm diameter × 50 μm long, frequently with characteristic pouch-like dorsal prolongations; cell division primarily in terminal region of filament; erect system reduced to bulb-shaped hair-bearing cells, seldom separated from prostrate system by further cell divisions; hairs sometimes in dense clusters, robust, brittle, up to 2.5 μm diameter, not sheathed, hairs exceptionally from cells of prostrate system; old hairs frequently golden brown at base; chloroplast of prostrate cells parietal with 1–10 (–12) conspicuous pyrenoids; chloroplast of hair-bearing cells generally plate-like with 1–2 (–3) pyrenoids.

Reproduction by quadriflagellate zoospores, up to 32–64 formed per cell in modified prostrate cells; zoospores 8.3–12 μm long, 3.4–6.5 μm diameter when settled, with chloroplast, eye-spot and single pyrenoid; biflagellate swarmers also reported.

Chromosome no.: n = 8 Moestrup (1969)
n = 11? Kermarrec (1970)

Plants endo- or epiphytic on a variety of loosely constructed species of brown and red algae, but especially in the tissues of *Chorda filum,* in littoral pools and in the sublittoral; tolerant of a wide range of conditions of exposure, depth and salinity (down to about 20‰).

Plants can be found all round the coasts of the British Isles from the Shetland Isles to the south coast.

Baltic Finland, Scandinavia south to Brittany, France; Helgoland.

North American Atlantic coast from Newfoundland south to New York; Texas; British Columbia, California; Arctic Sea, White Sea.

Plants can be found throughout the year in Britain; sporangia with zoospores were recorded in Anglesey, Wales, in February, June, July and November (South 1968), and in August in Dorset; sporangia were recorded in October in Brittany, France (Huber, 1892*a*), and in July in northern Norway (Jaasund, 1965); plants were recorded as present in the Baltic in summer with sporangia in autumn (Lakowitz, 1929). The life history is not completely known, but probably usually involves a succession of asexual generations; the function of biflagellate swarmers is not known. None of the workers who have grown *Bolbocoleon piliferum* in culture have found any evidence of meiosis in the life history. South (1968) found the quadriflagellate zoospores to germinate directly to give self-replicating generations. This was confirmed by Moestrup (1969) and Kermarrec (1970). Yarish (1975) also described and illustrated biflagellate swarmers, but was unable to determine the part they played in the life history. Biflagellate swarmers were also reported by Huber (1892*a*) and Hygen (1937).

The form variation of *Bolbocoleon piliferum* has been studied in culture by several workers with apparently conflicting results. South (1968, 1969), from culture experiments, found two forms of *Bolbocoleon:*
1 The most common 'normal' plants resembled plants seen in the field; they produced both hairs and sporangia.

2 'Reduced' plants with few hairs and with most cells developed into sporangia. He found that the reduced plants could not be associated with any particular time of the year, parent type or culture conditions. Yarish (1975) found a similar range of variation in morphology and growth. In 1976 Yarish reported that hairs were always formed under the ranges of nitrate and phosphate used in his culture media. Moestrup (1969), however, found the morphology of *Bolbocoleon* to be highly dependent on the culture medium he used. In 'poor' medium growth was slow; four days after transfer to the medium, hair cells appeared and hairs after eight days, but no sporangia were formed after two months. In 'rich' medium, growth was rapid and no hairs were formed; sporangia were formed after 28 days. The zoospores gave rise to plants of similar appearance.

Recorded hosts for *Bobocoleon piliferum: Chorda filum, Leathesia difformis, Mesogloia divaricata, Eudesme virescens, Dictyosiphon foeniculacea, Scytosiphon lomentaria, Chylocladia verticillata, Laminaria hyperborea, Dumontia contorta, Nemalion helminthoides, Laurencia obtusa, Polysiphonia violacea.*

ENTOCLADIA Reinke

ENTOCLADIA Reinke (1879), p. 476.

Type species: *E. viridis* Reinke (1879), p. 476.

Entoderma Wille (1890), p. 94.
Ectochaete (Huber) Wille (1911), p. 79.
Epicladia Reinke (1889), p. 31.

Thallus of prostrate, uniseriate, branched filaments, sometimes forming pseudo-parenchyma; erect system, when present, represented only by hairs; cells uninucleate, square or rectangular in surface view; hairs, when present, sometimes with bulbous bases; chloroplast parietal with one or more bilenticular pyrenoids; senescent cells not producing extra-cellular mucilage or sheaths.

Reproduction by biflagellate gametes, anisogamous or isogamous; also by quadriflagellate zoospores, in one species tri-flagellate zoospores; sporangial and gametangial mother cells with sequential cleavage; sporangia and gametangia cylindrical to flask-shaped with single exit papilla, rupturing explosively to liberate spores without surrounding vesicle.

An historical account of the genus *Entocladia* since it was first described by Reinke (1879) has been given by Yarish (1975). The name *Entocladia* disappeared subsequently as a result of the removal of the type species *E. viridis*, first to *Phaeophila* as *P. viridis* (Reinke) Burrows (Parke & Dixon, 1976) and then to *Acrochaete* as *A. viridis* (Reinke) Nielsen (Nielsen, 1979) following the results of experimental culture work by Yarish (1975) and from her own culture work where she established that the plants can produce hairs under certain environmental conditions and that the hairs sometimes have bulbous bases. Both of these changes in the taxonomic position of *E. viridis* have proved unsatisfactory. Several taxa were left without names since the name *Entocladia* could no longer be used and the inclusion of *E. viridis* in *Acrochaete* altered the circumscription of this genus as, although the hairs might be similar, *Acrochaete* has both prostrate and erect systems, while *Entocladia viridis* has only a prostrate system. It is also clear that the presence or absence and type of hair is not a sufficiently stable character to be used to separate these chaetophoracean genera. Recently many other characters derived from studies of the development of the thallus and its reproductive bodies, life histories, cytokinesis and ultrastructure have been found which may form a sounder basis for their separation. In the light of some of this work O'Kelly & Yarish (1980, 1981) have reinstated the genus *Entocladia* and have shown it to be very distinct from the genus *Phaeophila*. Their action has been followed here.

Following Yarish (1975) the genus *Epicladia* has been included in the genus *Entocladia*. When he described *Epicladia flustrae* found growing on *Flustra foliacea*, Reinke (1889) thought that the epizoic character of the plants was sufficient to separate the entity generically from similar morphological entities. Since then some authors (Batters, 1902; Taylor, 1937; Yarish, 1975) have accepted that algae with a morphology similar to that of Reinke's *Epicladia* on algal hosts, are identical with it and that there is no basis for separating it generically from *Entocladia*. On the basis of his experimental work, Yarish (1975) suggested that the name *Entocladia flustrae* (Reinke) Taylor should be recognised. This has been accepted here.

KEY TO SPECIES

1   Plants epi- or endophytic, epi- or endozoic, but not boring in shells   2
    Plants shell boring   4
2   Majority of cells with a single pyrenoid   3
    Majority of cells with 2 – 3 pyrenoids; plants with more or less simple
       filaments or sparsely branched forming pseudoparenchyma; usually
       on *Cladophora* or *Chaetomorpha* . . . . *E. leptochaete*
3   Plants endophytic in *Zostera* leaves, both inter- and intra-cellular . . . .
    . . . . , ... *E. perforans*
    Plants epi-endophytic in brown algae, especially *Elachista* spp. and
       Pilayella littoralis; filaments short and sparsely branched; cells
       average about 9 μm diameter . . . . *E. wittrockii*
    Plants epi-endozoic in *Flustra foliacea* and other Bryozoa; also in a variety
       of brown and red algae; filaments forming pseudoparenchyma with
       free filament ends; cells quadrate, rectangular or polygonal . . . . *E. flustrae*
    Plants endophytic in a variety of green, brown and red algae; filaments
       radiating from a centre, sometimes forming pseudoparenchyma; cell
       shape irregular, cells averaging about 6 μm diameter . . . . *E. viridis*
4   Plants with branching system of filaments penetrating shells of molluscs;
       no surface system of filaments . . . . *E. tenuis*
    NB. There is an *Entocladia*-like alga growing in mollusc shells, as yet
       unnamed which has both a surface and a penetrating system.

Entocladia flustrae (Reinke) Taylor (1937), p. 54.   Fig. 28

Lectotype: original illustration (Reinke, 1889, pl. 24) in the absence of material. Western Baltic, Germany.

Epicladia flustrae Reinke (1889), p. 31.

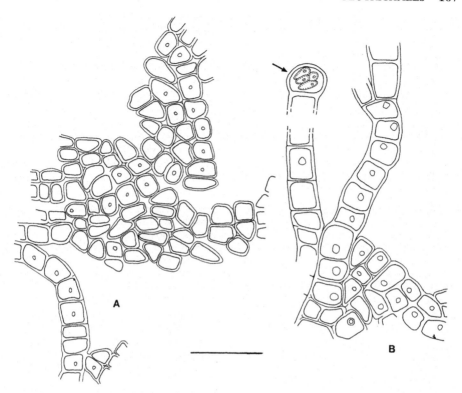

Fig. 28   *Entocladia flustrae*
A, B Portions of pseudoparenchymatous thalli. Note sporangial cell (arrowed) in B. A.
Thallus endozoic in *Sertularia*. (W. F. Farnham; April). B. (T. Norton; Nov.). Bar= 20 μm

Endoderma flustrae (Reinke) Batters (1902), p. 14.

Thallus consisting of branched filaments coalescing to form pseudoparenchyma with filament ends often free; cells rectangular or polygonal in surface view, 3–8 (–10) μm, up to six times long as broad in free filament ends, sometimes with hairs, bases of hairs swollen; chloroplast a lobed parietal plate; pyrenoid usually single.

Reproduction by quadriflagellate zoospores; zoosporangia clavate to flask-shaped, with single exit papilla, formed from any vegetative cell; zoospores 8–16 per sporangium, naked, pyriform, 3.5–5.0 × 6.5–9.0 μm, with cup-shaped chloroplast, one pyrenoid and stigma; zoospores released explosively with no enclosing vesicle; zoospores germinate without forming germ tube.

Chromosome no.: not known.

Plants grow on or in the exoskeleton of *Flustra foliacea*, *Sertularia* and other Bryozoa, also on various algae growing in the lower littoral and sublittoral regions;

whole stands of *Flustra foliacea* are sometimes turned green by this alga. Algal hosts reported include *Laurencia pinnatifida* and *Polysiphonia* species.

Reported by Batters (1902) as common on the coasts of England and southern Scotland, but recent records are few.

Kent, Dorset, Caernarvon, Glamorgan, East Lothian, Argyll, Co. Cork.

North European coast from Norway to France.

Greenland; Canadian Arctic south to New York, Florida; Texas; British Columbia.

Found all the year round in the British Isles, though winter records are relatively few; reproductive plants recorded for the summer months; plants are said to be present all the year round in the Baltic (Lakowitz, 1953), but in the Oslofjord the host animal is reported to be destitute of the epizoan (Sundene, 1953). Only sterile plants of *Entocladia flustrae* are known for eastern Canada (South, 1974), but fertile plants are recorded in July in Greenland (Lund, 1959). The life history appears to involve only a succession of asexual generations.

Hairs with bulbous bases are produced by cells of *Entocladia flustrae* under some conditions, especially those of a nitrate and phosphate deficiency (Yarish, 1976). On different hosts and on the same host at different tidal positions, the shape and size of cells of the plants show differences and there can also be physiological differences. It has been suggested that *E. flustrae* found in sublittoral hydroids, particularly in species of *Sertularia*, may represent an entity distinct from that found in sublittoral *Flustra foliacea* and *Securiflustra securifrons* (O'Kelly & Yarish, 1981). O'Kelly & Yarish suggest that the problem of the identity of the two forms needs comparative field and culture work.

Entocladia leptochaete (Huber) nov. comb.   Fig. 29

Lectotype: original illustration (Huber 1892, p. XV, figs 1–9) in the absence of material. Croisic, Finistère, France.

Endoderma leptochaete Huber (1892), p. 319.
Ectochaete leptochaete (Huber) Wille (1909), p. 79.
Phaeophila leptochaete (Huber) Nielsen (1972), p. 262.
Acrochaete leptochaete (Huber) Nielsen (1983), p. 69.

Plants epi- or endophytic, composed of microscopic branched filaments; filaments long with only occasional short branches or with many branches forming a pseudoparenchyma; cells 5–11 × 10–20 (–30) μm, sometimes with long hyaline hairs extending through the wall of the host; hairs slightly constricted above bulbous base, constriction sometimes with plug-like thickening; chloroplast parietal with irregular prolongations, fenestrate in older cells; pyrenoids 2–3 (–6), rarely single.

Reproduction by quadriflagellate zoospores 3.0–3.5 × 3.5–5.0 μm; also by biflagellate swarmers of two sizes, 2.0–2.5 μm long and 4.0–5.0 μm long.

Chromosome no.: not known.

Endophytic in outer walls of algae, particularly in species of *Cladophora* and *Chaetomorpha*; also reported in *Ceramium diaphanum*; found in the littoral and sublittoral regions.

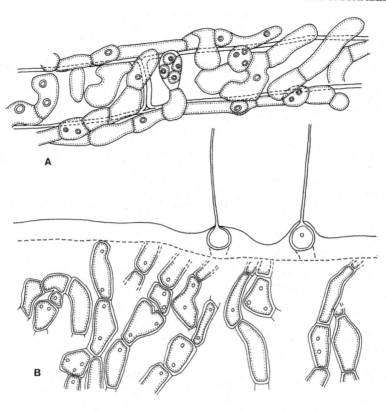

Fig. 29 *Entocladia leptochaete*
A. Thallus epiphytic on *Cladophora albida* (Kimmeridge, Dorset; July 1975). B. Thallus endophytic in *Cladophora* sp. (Loch Obiscens; H. Blackler; Dec. 1978). Bar= 20 μm.

Rarely recorded in the British Isles, but may be more common than the records suggest: Channel Islands, Dorset, Devon, North Uist.
Sweden south to France and the Mediterranean.
Not recorded outside Europe.

The seasonal behaviour of *Entocladia leptochaete* is not known but plants have been found during the summer months and in Dorset they have been found in a reproductive condition in July. The life history is not completely known. Nielsen (1983) found quadriflagellate zoospores which in her cultures grew into short filaments. She also found biflagellate swarmers of two sizes, but did not observe fusions between them.

There is a possibility that *Entocladia leptochaete* and *E. viridis* are conspecific. Comparing their own observations with those of Nielsen (1979), O'Kelly & Yarish (1980) decided that on the basis of vegetative cell morphology and dimensions, numbers of pyrenoids per cell, seta (= hair) production and morphology, habitat selection, morphology and dimensions of male gametes, it was not possible to separate the two species.

The present author has found plants answering to the original description of *E. leptochaete* given by Huber (1892*a*). The plants consistently have 2–3 pyrenoids per cell and rarely a single pyrenoid as in *E. viridis*; the branching pattern consists of elongated filaments of long cells with only short lateral branches here and there; hairs are frequent and usually have a bulbous base, sometimes with only a slight constriction. Morphologically these plants seem very distinct from *E. viridis* and while it is possible that they represent a variant of the latter species, the case does not seem yet proven. The view that they are distinct has also been taken by Nielsen (1983) who cultured the plants. However, she placed the taxon in *Acrochaete* as *A. leptochaete* (Huber) Nielsen mainly on the bulbous shape of the base of the hairs. But the plants lack the erect system of an *Acrochaete*.

Entocladia perforans (Huber) Levring (1937), p. 26.

Lectotype: original illustration (Huber, 1892, pl. XIV) in the absence of material. Lagoons, Golfe de Lion, France.

Endoderma perforans Huber (1892), p. 316.

Thallus of branched filaments radiating from a centre, ramifying both between and within cells of host; cells (3–) 4–10 (–14) μm diameter; hairs absent; chloroplast parietal with prolongations towards centre of cell, sometimes somewhat reticulate; pyrenoid usually single.

Reproduction by quadriflagellate zoospores; zoosporangia globular, formed from swollen vegetative cells, with papilla passing to exterior of host; zoospores 8–16 per sporangium, ovoid to almost spherical, 6–8 × 9.5–11 μm, with chloroplast covering posterior wall and with red eyespot.

Chromosome no.: not known.

Plants are found in old discoloured and floating leaves of species of *Zostera* in the lower littoral and sublittoral; they are said also to occur in colonies of *Rivularia* and other blue-green algae in the upper littoral region, though these records may refer to an entity other than *Entocladia perforans*.

The species has rarely been recorded in the British Isles. It is known to be present in the *Zostera* beds in Dorset from recent records and may also occur in other localities. Scandinavia south to the Atlantic coast of France; Baltic Sea.

On the North American Atlantic coast from Newfoundland south to Maine and Massachusetts.

Plants can be found forming swarmers in late summer when the leaves of *Zostera* plants are beginning to discolour with age and become detached and free-floating. Such leaves sometimes form deep piles at the edge of the tide and may persist well into the winter in Dorset.

The life history, as far as is known, involves a succession of asexual generations.

An alga identified as *E. perforans* has been described by Wilkinson (1969) and Wilkinson & Burrows (1972*a*), occurring on and boring into shells. The thallus is composed of two systems of branching filaments, a pseudoparenchymatous system on the host surface with rounded or isodiametric cells 4–12 μm in diameter, the cells sometimes in groups of 2, 3 or 4; the penetrating system, arising from the surface system, is of branched filaments with branching alternate or irregular in distal parts and opposite in older parts, cells sometimes swollen at the nodes. The chloroplast is parietal forming a larger or smaller ring and appearing to fill the cell completely in the boring filaments; there are 1–4 inconspicuous pyrenoids. A similar alga was found on chalk cliffs in Kent by Anand (1937) and identified by him as *E. perforans*.

In her studies of *E. perforans* isolated from *Zostera* leaves Nielsen (1980) recorded the number of pyrenoids as usually one but two in cells she thought were dividing. In culture experiments, Nielsen (loc. cit.) found that the filaments she grew on oyster shells and calcite crystal never penetrated into the calcareous substrate. Using material from the chalk cliffs studied by Anand (1937), she came to the conclusion that it belonged to *Pseudendoclonium submarinum*. There is at present no information suggesting that *Pseudendoclonium submarinum* is shell-boring and Nielsen (1980) suggested that Anand's penetrating filaments were merely growing between fragments of loose rock. In view of this, it would seem that more work is necessary to establish the identity of Wilkinson's (1969) plant. This is especially urgent as it is very common in many different species of shell all round the coast of the British Isles (Wilkinson & Burrows, 1972*b*).

Nielsen's (1980) combination *Epicladia perforans* has not been used since *Epicladia* was merged with *Entocladia* following Yarish (1975).

Entocladia tenuis Kylin (1935), p. 201.   Fig. 30

Lectotype: original illustration (Kylin, 1935, p. 17. fig. 7) in the absence of material. Kristineberg, Sweden.

Phaeophila tenuis (Kylin) Nielsen (1972), p. 258.

Thallus endozoic, of branched cellular filaments originating from a single cell on host surface, lacking external cell system; branching irregular, occasionally several branches from one cell; hairs sometimes present, 1 per cell, 1–2 μm diameter at base, gradually tapering to apex, up to 100 μm long, hollow for most of length, closed at base; cells with slightly undulate lateral walls, 3–12 μm broad, 4–10(–20) × long as broad in younger thallus 4–8 × long as broad in older parts; chloroplast parietal, lobed, occupying less than length and width of cell; 1–4 pyrenoids per cell.

Reproduction by quadriflagellate zoospores formed in slightly enlarged thallus cells at any point in thallus; zoospores 6–8 × 4 μm.

Chromosome no.: not known.

Filaments endozoic in mollusc shells in the shallow sublittoral and cast up on tidal flats.

Only fairly recently recorded for the British Isles by Wilkinson (1975): Ayrshire, East Lothian.

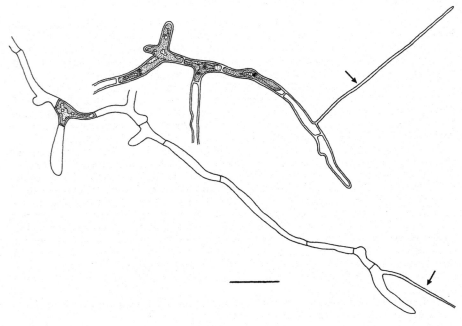

Fig. 30 *Entocladia tenuis*
Branched endophytic filaments with hairs (arrowed). Figure drawn by M. Wilkinson.
Bar= 50 μm.

Norway, Sweden, Denmark. The species does not appear to be as common in the British Isles as in Scandinavia.
East coast of USA.

The plants of *Entocladia tenuis* from Ayrshire and East Lothian were collected by Wilkinson (1975) in April 1972 and were apparently not then reproducing. In Denmark plants with sporangia were collected by Nielsen (1972) in June 1969 and August 1970. Kylin (1935) also reported reproductive plants during the summer. Only quadriflagellate zoospores have so far been recorded and the life history appears to be a succession of asexual generations.

The presence of hairs is not a constant feature of this species. The original description of Kylin (1935) gave it as a hairless plant and all records except those from Denmark and Britain are of hairless plants. Nielsen (1972) has shown that in culture the presence of hairs can be modified by the environment. There are many unidentified taxa present in small amounts in British shells and it is possible that some of these may be unrecognised hairless forms of *Entocladia tenuis*.

Entocladia viridis Reinke (1879), p. 476.   Fig. 31

Lectotype: original illustration (Reinke, 1879, fig. 6) in the absence of material. Naples, Italy.

Endoderma viride (Reinke) Wille (1890), p. 94.
Phaeophila viridis (Reinke) Burrows in Parke & Dixon (1976), p. 568.
Acrochaete viridis (Reinke) Nielsen (1979), p. 442.

Thallus endophytic, of microscopic branched filaments with apical and intercalary division, radiating from a centre, rarely forming pseudoparenchyma; cells with or without hairs; hairs with or without swollen bases, separated by a wall from the hair-bearing cell; cells cylindrical, rounded or irregular in shape, 3–10 μm broad, 1–2 (–6) times as long as broad; chloroplast a parietal plate, lobed with usually a single bilenticular pyrenoid, occasionally up to 4 per cell.

Reproduction by biflagellate and quadriflagellate swarmers; biflagellate swarmers may act as gametes; gametangia and zoosporangia formed from any vegetative cell, flask-shaped with exit tube; zoosporangia with 8–16 zoospores; zoospores 3.5–4.5 × 6.0–7.0 μm with parietal cup-shaped chloroplast and single pyrenoid and eyespot; biflagellate swarmers of two types, larger 2.8–3.8 (–5.0) × 4.5–6.0 (–7.0) μm with cup-shaped chloroplast, pyrenoid and eyespot, smaller 1.5–3.0 × 2.5–4.0 μm, colourless, eyespot absent.

Chromosome no.: n   = 8 O'Kelly & Yarish (1981)
                 2n = 16 ± 1

Plants endophytic in a variety of green, brown and red algae; also occurring in shells, found in the mid-lower littoral and sublittoral regions.

Generally distributed round the coast of the British Isles, but probably often missed and therefore more common than the records show. There are few records for Scotland. Channel Islands, Kent, Dorset, Devon, Cornwall, Lundy Island, Glamorgan, Isle of Man, Isle of Cumbrae, Isle of Arran, Tyne & Wear, Island of Colonsay, Co. Wexford, Co. Cork.

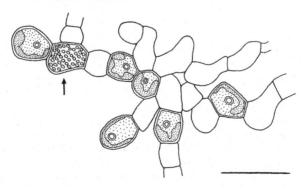

Fig. 31   *Entocladia viridus*
         Endophytic branched filaments in *Polysiphonia elongata*. Note reproductive cell with enclosed swarmers (arrowed). Bar= 20 μm.

North European coast from Scandinavia south to the Mediterranean and Adriatic Sea, Baltic Sea.

North American Atlantic coast from Newfoundland south to Massachusetts, North Carolina, Florida, Bermuda, Brazil; Pacific coast of North America from southern British Columbia to California, Galapagos Islands; New Zealand.

Plants can be found throughout the year; they reproduce during the summer months. The life history is not completely known. For isolates of *E. viridis* collected in Connecticut, USA, O'Kelly & Yarish (1981) found that the two types of biflagellate swarmers behaved like gametes and they completed an isomorphic alternation of gametophyte and sporophyte generations in culture. Nielsen (1979) also saw quadriflagellate zoospores and two sizes of biflagellate swarmers in cultures started from plants collected at Naples. There may, therefore be an isomorphic alternation of gametophyte and sporophyte generations also in Europe, but at no time did Nielsen see fusions between the two types of biflagellate swarmers. They developed directly into new plants. In the experiments by O'Kelly & Yarish (1981) the gametes did not develop parthenogenetically. So far meiosis has not been observed in the life history, but O'Kelly & Yarish (loc. cit.) found a chromosome count of 8 in sporangia, gametangia and gametophytic vegetative cells, while vegetative cells of the sporophyte had a count of 16 ± 1. They assumed that meiosis occurred at the first division of the sporangium. However, for an asexually reproducing isolate, they found all the nuclear divisions in the sporangium to be mitotic, with a count of 15 ± 2 (O'Kelly, 1980). The life history needs further study both in Europe and America. It was suggested by Kylin (1938) and reiterated by O'Kelly & Yarish (1981) that *E. viridis* is widespread and may represent an assemblage of taxa; this may be supported by the fact that cell size and shape can differ markedly between samples, though the different host plants may also play a part in this. For this reason O'Kelly & Yarish (loc. cit.) express some doubt as to whether all the distributional records refer to the same taxon.

Nielsen (1979) found that in cultures in contact with a firm substrate the plants grew more or less circular as pseudoparenchymatous cushions, with later branches growing freely into the medium from the cushions. Unattached thalli appeared as masses of uniseriate filaments with branches entangled centrally and with free branches at the periphery. Cell size and shape vary considerably between plants in different hosts; hairs tend to be formed more often and in greater numbers on plants growing in water deficient in nitrate and phosphate (Yarish, 1976).

Entocladia wittrockii Wille (1880), p. 3.

Lectotype: original illustration (Wille, 1880), in the absence of material. Oslofjord, Norway.

Phaeophila wittrockii (Wille) Nielsen (1972), p. 264.
Ectochaete wittrockii (Wille) Kylin (1938), p. 72.

Thallus epiphytic or endophytic in cell wall of host, composed of relatively short filaments of up to about 10 cells in length, simple or sparsely branched, tapering at ends; cells 4–5 (–10) μm broad, (1–) 2–3 times as long as broad; cells with conspicuous hairs without basal swelling; chloroplast parietal, sometimes dissected or perforated, with a single bilenticular pyrenoid.

Reproduction by triflagellate, sometimes quadriflagellate zoospores formed in cylindrical to flask-shaped sporangia with exit papilla, formed from vegetative cells; cleavage of sporangia sequential forming 4–64 zoospores; zoospores pyriform, naked 4.8–6.5 × 5.0–7.5 μm, with cup-shaped chloroplast, one pyrenoid and two red eyespots; zoospores released through ruptured exit papilla without enclosing vesicle.

Chromosome no.: 25–40 O'Kelly & Yarish (1981).

The species occurs in or on the cell walls of brown algae, less commonly in red algae; found in the littoral region and in the sublittoral down to 14–15 m, in both sheltered and exposed localities; frequent in areas where salinity is reduced or fluctuating.

Recorded hosts; *Pilayella littoralis, Ectocarpus siliculosus, Elachista fucicola, E. scutulata, Sphacelaria arcta, Striaria attenuata, Desmotrichum undulatum, Leptonema fasciculatum, Dictyosiphon chordaria, Polysiphonia* species.

There are relatively few records for this species in the British Isles and records listed as *Entocladia wittrockii* may not, in fact, belong to this species. On the other hand, the species may be more common than is indicated by the records. Newton (1931), as has been pointed out by Levring (1937) confused the species with *E. leptochaete*, the figure of this latter species being taken from Wille's original description of *E. wittrockii*.

Kent, Glamorgan, N. Wales, Isle of Man, S.W. Scotland, Northumberland.

Scandinavia south to the Atlantic coast of France, Baltic Sea.

Greenland, Atlantic coast of North America from Newfoundland south to Maine, Massachusetts, Gulf of Mexico, Florida; British Columbia on the Pacific coast of North America.

The species is reported to occur and to reproduce at all times of the year (Knight & Parke, 1931). In Norway it appears first in spring, becoming fertile from May to September or October, and then declines (Printz, 1926). In eastern Canada plants have been recorded in most months of the year, though most commonly in summer and autumn and are fertile mainly in summer (South, 1974). The life history is not completely known, but appears to involve only an asexual sequence of generations.

A number of authors have described triflagellated swarmers for *E. wittrockii* (Kylin, 1938; Kornmann, 1959; Schussnig, 1960; South, 1974; O'Kelly & Yarish, 1981; Nielsen, 1983). O'Kelly & Yarish (loc. cit.) saw no other type of swarmer and these were reported to develop into filaments similar to the parent filaments and again to form triflagellate swarmers. Kornmann (1959) recorded both tri- and quadriflagellate swarmers in cultures, the quadriflagellate swarmers being formed later than the triflagellate. The triflagellate swarmers could develop into single-celled sporangia which in turn produced more triflagellate swarmers. He also found a possible sexual process involving biflagellate swarmers. Nielsen (1983) found all three types of swarmer in her cultures, including two sizes of biflagellate swarmers which she thought might be anisogametes, but she saw no fusions between them. Meiosis has not been seen in any of the work described. O'Kelly & Yarish (1981) found only mitotic divisions within the sporangium in their cultures. They (1981, loc. cit.) have postulated that the high chromosome count (25–40) represents a polyploid condition and suggest that this may explain the aberrant structural and reproductive features of the species.

There is uncertainty about the formation of hairs in *E. wittrockii*. They were not mentioned by Wille (1880) who first described the species, but they have since been recorded by Kylin (1938), O'Kelly & Yarish (1981) and Nielsen (1983). O'Kelly & Yarish found hairs only in some of the plants they collected from the Pacific coast and they described them as generally lacking a swollen base. South (1974) recorded hairs as present or absent for this species in Newfoundland and stated that the hairless variety had a more superficial growth on the host plant. Nielsen (1983) has recorded hairs of the *Acrochaete* type, i.e. swollen at the base, on plants in her cultures that had been kept for a week in unenriched seawater.

OCHLOCHAETE Thwaites in Harvey

OCHLOCHAETE Thwaites in Harvey (1849), pl. CCXXVI.

Type species: *O. hystrix* Thwaites in Harvey (1849), pl. CCXXVI.

Chaetobolus Rosenvinge (1893), p. 928.

Thallus a more or less pseudoparenchymatous disc of irregular branched filaments radiating from a central point, initially monostromatic, becoming 2–3 layered, often surrounded by a gelatinous sheath; cells with long hyaline unsheathed hairs tapering to distal end; chloroplast parietal with prolongations towards cell centre; pyrenoid 1, occasionally 2–3.

Reproduction by quadriflagellate zoospores formed in central disc cells.

Three species of *Ochlochaete* have been described and a fourth taxon *Chaetobolus gibbus* appears to belong to the same complex. *Ochlochaete hystrix* was the first species to be described (Thwaites in Harvey, 1849). Two further species of *Ochlochaete*, *O. ferox* and *O. lentiformis* were described by Huber in 1892. *O. lentiformis* was found growing on pieces of glass, pottery and 'old pipes' rather than epiphytic as were the other two species. Otherwise the descriptions of the three species differed only in terms of the regularity or irregularity of the disc, its compactness, the size of the cells and the frequency of occurrence of hairs and by the stage in development at which the thallus becomes polystromatic. This last character separated the three species of *Ochlochaete* from Rosenvinge's species *Chaetobolus gibbus*. Recent experimental work by Yarish (1975, 1976) has shown that all of these characters are modified by the environment and he suggested that all four species might be conspecific. Nielsen's (1977) work on *O. hystrix* and *O. ferox* also supported this conclusion: as a result she placed *O. ferox* as a variety of *O. hystrix*. Also as a result of her experimental work, she thought that *O. lentiformis* and *Ulvella lens* are conspecific, having shown that the latter species could produce hairs under some culture conditions. For the purposes of this volume, the three species of *Ochlochaete* have been accepted as conspecific and distinct from *Ulvella lens*, but it is clear that there is still a problem here. The genera *Ochlochaete*, *Chaetobolus* and *Ulvella* are close to one another morphologically and the question of their relationship still needs clarification.

The fact that long colourless hairs have been recorded in plants of *Pringsheimiella* collected from the field (Reinke, 1888; Printz, 1926; Newton, 1931; Nielsen & Pedersen, 1977) brings this genus close to the *Ochlochaete/Ulvella* complex: this also needs further consideration.

Only one species in Britain.

Ochlochaete hystrix Thwaites in Harvey (1849), pl. CCXXVI.

Lectotype: original illustration (Harvey, pl. CCXXVI) in the absence of material. 'The little sea', Studland, Dorset.

Ochlochaete ferox Huber (1892), p. 355.
Ochlochaete lentiformis Huber (1892), p. 355.
Chaetobolus gibbus Rosenvinge (1893), p. 928.

Thallus of branched filaments radiating from centre, forming regular or irregular disc, 50–800 μm diameter, filaments adhering laterally; disc sooner or later becoming polystromatic; branching of filaments lateral, sometimes becoming dichotomous at margins of disc; cells spherical or oblong in centre of disc, distinctly longer than broad towards margin; marginal cells 3.5–10.5 × 29–50.5 μm; cells, sometimes only central cells, each with single long rigid hair ending in a fine point; hairs not cut off from mother cell; chloroplast parietal, lobed or irregular, sometimes with prolongations towards cell centre; pyrenoid usually single, rarely up to 3 or more.

Reproduction by quadriflagellate zoospores, occasionally also by biflagellate swarmers; zoosporangia formed from swollen central disc cells, flask-shaped with long neck; zoospores 16–64 per sporangium, released through the neck surrounded by a mucilage envelope; zoospores with hyaline beak, naked, spherical to ovoid, 3.5–6.5 × 4.5–9.5 μm, with chloroplast, 1–3 pyrenoids and eyespot; biflagellate swarmers not described.

Chromosome no.: not known.

Plants are found in both brackish and fully saline conditions, epiphytic on a variety of algae and aquatic angiosperms, also epilithic and epizoic in the littoral and shallow sublittoral regions.

Probably more common than is indicated by the records.
Dorset, Gloucestershire, Isle of Cumbrae, Shetland Isles.
European Atlantic coast from Scandinavia south to France, Baltic Sea.
Atlantic coast of North America from Newfoundland south to Massachusetts; Greenland; Iceland.

In the British Isles probably present throughout the year, but in more northerly waters found only during the summer months: swarmers formed in summer. The life history is incompletely known: there is a succession of asexual generations, but the fate of the biflagellate swarmers is not known; some observations suggest that they behave as zoospores (Nielsen, 1978).

The plants show a range of variation in cell size, branching pattern of the thallus and occurrence of hairs. Yarish (1976) on the basis of culture experiments decided that the variation in these characters is environmentally produced and responsible for the differences shown by the described species, *Ochlochaete hystrix* Thwaites in Harvey (1849), *O. ferox* Huber (1892), *O. lentiformis* Huber (1892) and *Chaetobolus gibbus* Rosenvinge (1893). His conclusions are supported, for *O. hystrix* and *O. ferox*, by the work of Nielsen (1977).

PHAEOPHILA Hauck

PHAEOPHILA Hauck (1876), p. 56.

Type species: *P. floridearum* Hauck (1876), p. 56 (= *P. dendroides* (Crouan frat.) Batters (1902), p. 13).

Thallus epi- or endophytic, of free prostrate irregularly branched filaments; cells with long undulate, sometimes wrinkled hairs, only occasionally cut off by cross walls; chloroplast parietal, lobed with usually several pyrenoids; chloroplast sometimes appearing reticulate.

Reproduction by quadriflagellate zoospores formed in slightly, sometimes irregularly swollen, but never flask-shaped sporangia; mother cells multinucleate, cleavage simultaneous; zoospores released through nonexplosive opening of neck; extra cytoplasm in sporangium cavity extruded as a plug through the neck.

It has been pointed out by O'Kelly & Yarish (1980) that *Phaeophila* is unlike the other marine members of the Chaetophoraceae and shows similarities to *Blastophysa* Reinke. *Blastophysa* undergoes synchronous mitoses in the sporangium mother cell without attendant cytokinesis (Sears, 1967) and has a sporangium neck that opens well before release of the zoospores (Huber 1892a). The mitotic figures and the structure of the zoospores of both genera are remarkably alike. O'Kelly & Yarish (loc. cit.) saw *Phaeophila* filaments in their cultures with a *Blastophysa*-like structure. They suggested that *Phaeophila* might be better placed along with *Blastophysa* in the Chaetosiphonaceae than among the Chaetophoraceae.

Only one species.

Phaeophila dendroides (Crouan frat.) Batters (1902), p. 13. Fig. 32

Type: CO. Rade de Brest, Finistère, France.

Ochlochaete dendroides Crouan frat. (1867), p. 128.
Phaeophila floridearum Hauck (1876), p. 56.
Ochlochaete phaeophila Falkenberg (1879), p. 233.
Phaeophila engleri Reinke (1889), p. 86.

Thallus endophytic or endozoic, composed or irregularly branched filaments, rarely forming pseudoparenchyma; cells elongate, sometimes swollen, 8–40 μm diameter, 1–4 (–6) times as long as broad, often bearing long undulate and twisted hairs; hairs only occasionally cut off by cross walls; chloroplast parietal, often unevenly thickened so as to appear finely reticulate; pyrenoids one to several.

Reproduction by quadriflagellate zoospores; zoosporangia formed from swollen vegetative cells, rounded with long neck, hyaline plug at neck apex; divisions in zoosporangium simultaneous to form 16–49 zoospores; zoospores pear-shaped, 5–8 × 12–14 μm with 1 pyrenoid and eyespot, released through neck of sporangium without enclosing vesicle; biflagellate swarmers are also reported (Hauck, 1876; Huber, 1892a).

Chromosome no.: n = 17 ± 2 O'Kelly & Yarish (1980).

Endophytic in the outer membrane or between cortical cells of larger algae and endozoic in mollusc shells in both littoral and sublittoral region.

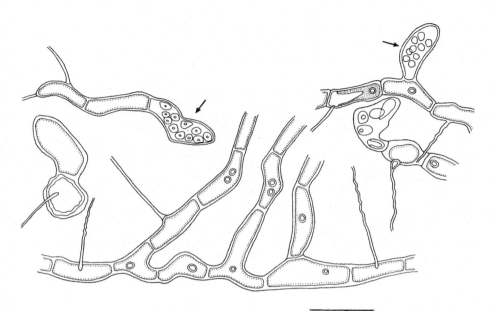

Fig. 32   *Phaeophila dendroides*
Portions of branched filaments in *Polysiphonia elongata* with long, thin, thread-like hairs and reproductive cells (arrowed). (Lough Ine, Co. Cork; 1959). Bar= 50 μm.

Reported hosts: *Spirorbis* species, *Balanus balanoides, Cardium edule, Littorina littorea, Mya arenaria, Solieria chordalis, Polysiphonia elongata, Chorda filum, Laminaria digitata.*

Probably more frequent and widespread than is recorded.
Kent, Dorset, Isle of Man, Isle of Colonsay, Co. Wexford, Co. Cork.
On the Atlantic coast of Europe from Scandinavia south to France; Mediterranean.
Eastern Canada, Bermuda; British Columbia, Washington.

Has been recorded sporadically through the year; zoospores formed during the summer months. The life history of *Phaeophila dendroides* is not completely known. Field records refer to the formation of quadriflagellate zoospores and O'Kelly & Yarish (1980) found only mitotic divisions in the sporangium. From this it might be assumed that the life history involves only a succession of asexual generations. Huber (1892a) described biflagellate swarmers also, released from plants growing in *Zostera* leaves. So far no meiosis has been seen and the part played by the biflagellate swarmers in the life history is not known.

The cells vary in size and shape, but the long hairs, which are frequently twisted especially towards the base and which lack a basal swelling, are characteristic for this species.

Phaeophila dendroides was first described as *Ochlochaete dendroides* by the Crouan brothers in 1859 and the type material on *Solieria chordalis* was, according to Nielsen (1972), distributed as No. 346 in their Exsiccata *Algues Marines du Finistère*. Nielsen checked the specimen kept at the Botanical Museum of Copenhagen and gives a figure (1972, p. 256).

Phaeophila floridearum was described as the type species of the new genus *Phaeophila* by Hauck in 1876. Huber (1892a) commented that *Ochlochaete dendroides* Crouan frat. was very similar to *Phaeophila floridearum* Hauck though the former had hairs that were not twisted while in the latter they were twisted. Huber suggested that *O. dendroides* should be placed in *Phaeophila* Hauck., but this was not done until 1902 when Batters listed the name *Phaeophila* (= *Ochlochaete*) *dendroides* Batt.

Phaeophila engleri was said by Reinke (1889), who described it, to differ from *P. dendroides* by its shell-boring habit and its swollen cells. Thivy (1943) stated that it differed from *P. dendroides* also by having much shorter exit tubes from the sporangia. Nielsen (1972), in experiments with plants extracted from shells, found that they would grow quite well in culture. The cells then were straight and not swollen as they were in the shell, indicating that cell shape varies with substrate. Her cultures also showed very long exit tubes from the sporangia. Comparing her cultured plants with type material of *Phaeophila dendroides*, she decided that *P. engleri* and *P. dendroides* are conspecific.

PIRULA Snow

PIRULA Snow (1911), p. 363.

Heterogonium Dangeard (1912), p. XVII nom. illeg., *non Heterogonium* Presl (1849), p. 502.

Plants unicellular, solitary, in chains of up to 3 cells, or grouped in colonies of 3–4 cells; chloroplast parietal, pyrenoid single.
Reproduction by budding.
The genus has both freshwater and marine species.

Only one marine species recorded.

Pirula salina (Dangeard) Printz (1927), p. 225.

Type: BORD. In seawater culture vessel, Bordeaux, France.

Heterogonium salinum Dangeard (1912), p. XVII.

Cells bright green, oval, 9–10 (5–7) × 5–6 (3–5) μm; cell membrane colours blue with iodide.
Reproduction by budding; cells usually separated before bud has reached dimensions of parent, but sometimes remaining together.

Chromosome no.: not known.

Found in seawater and in microflora of estuary installations.

Plymouth? No material to substantiate this record can be found (Boalch and Green, pers. comm.); the species may have been regarded as planktonic by Parke & Dixon (1976) and included in the third revision of the Check-list on the basis of records from the other side of the Channel (see Gayral & Lepailleur, 1968). France.

PRINGSHEIMIELLA Hohnel

PRINGSHEIMIELLA Hohnel (1920), p. 97.

Type species: *P. scutata* (Reinke) Marchewianka (1924), p. 42 (as *P. scuttata* (sic!)).

Pringsheimia Reinke (1888), p. 241, nom. illeg. *non Pringsheimia* Schulzer (1866), p. 57 nec *Pringsheimia* Wood (1872), p. 195.
Syncoryne Nielsen & Pedersen (1977), p. 415.

Thallus a monostromatic, pseudoparenchymatous disc of radiating branched filaments with marginal division, sometimes surrounded by gelatinous membrane at least in young stages; central cells with longest diameter vertical to basal plane, grading to flatter cells, radially elongated on periphery; chloroplast a parietal disc or plate with one pyrenoid; cells sometimes with long colourless hairs.

Reproduction by biflagellate gametes, biflagellate zoospores, or quadriflagellate zoospores (see Nielsen & Pedersen, 1977).

The occasional occurrence of long colourless hairs on plants of *Pringsheimiella* collected from the field has been reported by a number of authors (Printz, 1926; Newton, 1931; Nielsen & Pedersen, 1977) and brings this genus close to the genus *Ochlochaete*. Yarish (1976), in culture experiments with *Ochlochaete hystrix*, found that when the plants were grown at a temperature of 32° C, regardless of light

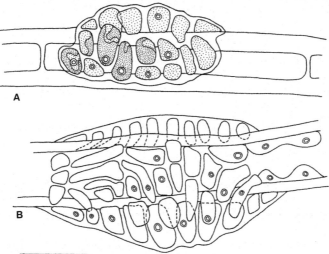

Fig. 33   *Pringsheimiella scutata*
A, B. Pseudoparenchymatous discs epiphytic on *Rhizoclonium riparium*. Bar= 20 μm.

intensity, they bore a remarkable resemblance to *Pringsheimiella scutata* (see discussion on the genus *Ochlochaete* p. 116). Recently Nielsen & Pedersen (1977) created a new genus, *Syncoryne*, for a plant included as part of the life history of *Pringsheimiella scutata* Reinke (1888) in the original description of this species (see under *Pringsheimiella scutata* p. 123).

Only one species in the British Isles.

Pringsheimiella scutata (Reinke) Marchewianka (1924), p. 42.   Fig. 33

Lectotype: original illustration (Reinke, 1889, pl. 25) in the absence of material (see Nielsen & Pedersen, 1977). Kieler Förde, Germany.

Pringsheimia scutata Reinke (1888), p. 241.
?*Syncoryne reinkei* Nielsen & Pedersen (1977), p. 415.

Thallus composed of branched filaments radiating from a centre, forming a monostromatic, pseudoparenchymatous cushion-like disc; cells with outer walls more thickened than radial walls; outer walls sometimes striated; central cells wedge-shaped or isodiametric, up to 12 μm diameter, with long axis at right angles to base of disc; peripheral cells flatter and radially elongate, 4–10 μm diameter, 12–26 μm long, 8–10 μm high; cell divisions marginal, dichotomous; cells sometimes with hairs; hairs with collar at base; chloroplast parietal with single pyrenoid.

Reproduction by biflagellate swarmers which may behave as gametes or zoospores, 16–32 or more per cell, formed first from central disc cells, later from all cells; swarmers of two sizes smaller 2.4–3.0 (–3.6) × 3.6–4.2 (–4.8), larger 4.2–4.8 × 5.6–7.2; swarmers escape enclosed in a mucilage vesicle, through a short exit tube; reproduction also by quadriflagellate zoospores formed in cells of separate thalli; zoospores pyriform, 4.5–5.5 × 6.5–7.4.

Chromosome no.: n = 4 Nielsen & Pedersen (1977).

Plants occur as epiphytes on a wide range of algae in the lower littoral and shallow sublittoral regions in fairly sheltered locations; also on leaves of *Zostera*.

Widely distributed round the coast of the British Isles: Kent, Dorset, Lundy Island, Isle of Man, Hilbre Island (Cheshire), Isle of Cumbrae, Summer Isles (Ross), Shetland Isles, Tyne & Wear, Belfast Lough.
European Atlantic coast from Norway south to France, Mediterranean.
North American Atlantic coast from Labrador south to Florida and the Caribbean, Greenland; Iceland, Faroes; Virgin Isles.

Plants are present throughout the year, forming swarmers in late summer and autumn. There is evidence (Printz, 1926) that plants may overwinter even under arctic ice. The life history appears to be an alternation between gametophyte and sporophyte generations, but this may be a simplification as two distinct taxa may be involved. So far meiosis has not been observed.

The description given for *Pringsheimiella scutata* above is based on that for the 'ungeschlechtliche' plants of Reinke (1889a) and on that given for morphologically similar plants by Nielsen & Pedersen (1977). Reinke described the plants as forming biflagellate neutral swarmers while Nielsen & Pedersen found them to produce two sizes of biflagellate swarmers that acted as gametes, but which they thought might also act as zoospores. In their cultures they also found plants they referred to as 'lumpy' with cells arranged randomly rather than in a disc. These formed quadriflagellate zoospores which germinated to reform the disc-shaped plants.

Reinke (1889) described a second type of thallus which he called 'Geschlechtspflanze', the discs of which were composed of rounded cells whose walls were equally thickened all round; he described and illustrated the plants as producing biflagellate gametes, the fusion of which formed almost spherical zygotes. Plants of this same morphology were described by Nielsen & Pedersen (1977) as producing claviform sporangia from which quadriflagellate zoospores were released enclosed in a mucilage envelope. In culture the zoospores developed into new plants similar to the parents. At no time did they produce gametes. As had earlier been suggested by Kylin (1949), Nielsen & Pedersen (loc. cit.) regarded this second type of plant as having no relationship with that described above as *Pringsheimiella scutata*. They created a new genus, *Syncoryne*, for the species which they named *S. reinkei* Nielsen & Pedersen, typified by material collected at Århus in 1975 and kept at the Botanical Museum, Copenhagen. The two plant types have so far not been distinguished in Britain*. They apparently both have the same chromosome number of n = 4. The two forms are obviously closely related morphologically and though each can reproduce independently this does not rule out the possibility that they may belong to a single life history. Until the zygotes have been germinated and the form of the next generation determined, the relationship between them must remain an open question.

PSEUDENDOCLONIUM Wille

PSEUDENDOCLONIUM Wille (1901), p. 29.

Type species: *Pseudendoclonium submarinum* Wille (1901), p. 29.

Plants freshwater and marine, forming crusts of an irregularly branched, prostrate, often pseudoparenchymatous system of cells giving rise below to 1–, or rarely more than one-celled rhizoids, above to erect short irregularly branched filaments; branches often arising from middle of cells; hairs lacking; sometimes forming *Pleurococcus*-like colonies of cruciately divided cells; chloroplast parietal with one pyrenoid.

Reproduction by quadriflagellate zoospores, also by biflagellate swarmers and by akinetes.

Several freshwater species have been described for this genus but until now only one marine species has been recorded. Recently, however, Nielsen (1980) has transferred *Pseudopringsheimia fucicola* (Rosenvinge) Wille to *Pseudendoclonium* as *Pseudendoclonium fucicola* (Rosenvinge) Nielsen, on the grounds that its morphology and mode of fragmentation in culture resemble *Pseudendoclonium submarinum* and that it is unlikely to have the heteromorphic life history shown by Perrot (1969) for *Pseudopringsheimia confluens*, the second species in the genus *Pseudopringsheimia*. The life history of *Pseudopringsheimia fucicola* is still not completely known, nor is that of *Pseudendoclonium submarinum*.

One marine species treated here.

---

* material from the Thames Estuary identified as *S. reinkei* by Nielsen ; see microscope slide BM 12538 eds.

Pseudendoclonium submarinum Wille (1901), p. 29.   Fig. 34

Type: O. Drøbak, Norway.

Pseudendoclonium marinum (Reinke) Aleem & Schültz (1952), p. 71, 72.

Thallus composed of branched filaments associated to form a pseudoparechymatous layer over the substrate; erect filaments of 3–8 cells, irregularly branched, running down into rhizoid-like filaments forming basal layer; cells 5–10 μm diameter, (1–) 2–4 (–8) times as long as broad; hairs lacking; chloroplast a dense parietal plate, sometimes only partially covering cell wall; pyrenoid single, large; thallus sometimes separating into single-celled units.

Reproduction by quadriflagellate zoospores, 16–32, in oval sporangia; zoospores 3–5 × 4–8 μm, released through hole in apex of sporangium, enclosed in common mucilage envelope; biflagellate swarmers of two sizes sometimes occur; large 3.5–4.5 × 6.0–7.5 μm, smaller 2.0–3.5 × 4.0–5.0 μm; reproduction also by aplanospores and akinetes.

Chromosome no.: not known.

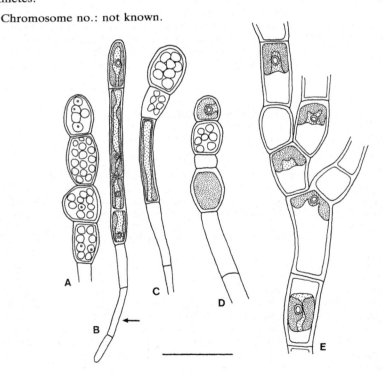

Fig. 34   *Pseudendoclonium submarinum*
A-E, Portions of erect simple and branched filaments. Note some cells reproductive with enclosed swarmers, and basal rhizoidal filaments (arrowed). (Severn Estuary; Feb. 1977; top shore rocks, Studland, Dorset; March 1976; L. Smith). Bar= 20 μm.

Forming crusts on consolidated mud, on and in woodwork, on rock, stones and shells in the upper littoral region; frequent in estuaries and on the edges of salt marshes. According to Jaasund (1965) the rhizoids fill up tracheids in woodwork.

Widely distributed in the British Isles, but probably often overlooked. Reported from: Kent, Dorset, Severn Estuary, Island of Cumbrae, Isles of Harris and Lewis, Co. Wexford, Co. Clare.

On the European Atlantic coast from Finmark, Norway, south to Spain and Portugal.

Atlantic coast of North America from Newfoundland south to Florida; Bermuda; Greenland.

Plants are present throughout the year; sporangia have been reported in September and October, but the full seasonal period of reproduction is not known. The life history is also not yet fully known. Quadriflagellate zoospores and two sizes of biflagellate swarmers have been reported but fusions between the latter have not been seen. According to Collins (1909) akinetes of two types are formed, one type germinating immediately and a thick-walled type which acts as a resting stage.

Pseudendoclonium submarinum was first described by Wille (1901). Aleem & Schulz (1952), however, decided that *Protoderma marina* Reinke and *Pseudendoclonium submarinum* Wille were sufficiently alike to be combined under the name *Pseudendoclonium marinum* (Reinke) Aleem & Schulz. Collins (1909) had described the two entities as separate species as did also Humm & Taylor (1961). *Pseudendoclonium* was described by Wille (1901) as pseudoparenchymatous and composed of closely packed short branched filaments with short rhizoids below. The description of *Protoderma marina* Reinke is less definite, but Collins (1909) described the filaments as 'parenchymatously united in the interior of the disc, free at the margin'. The genus *Protoderma* Kützing emend. Borzi (1895) was reported to have biflagellate zoospores with an eyespot and one pyrenoid, while Wille's *Pseudendoclonium* was described as having quadriflagellate zoospores with one pyrenoid, but lacking an eyespot. Lund (1959) was of the opinion that since the original description of *Protoderma marina* was incomplete, Wille's *Pseudendoclonium submarinum* should be retained. There is a good case for this and it has been followed here.

PSEUDOPRINGSHEIMIA Wille

PSEUDOPRINGSHEIMIA Wille (1909b), p. 88.

Type species: *Pseudopringsheimia confluens* (Rosenvinge) Wille. (See South, 1974, p. 919).

Thallus cushion-like with both prostrate and erect systems of branched filaments; filaments coalescing to form a closely compact cushion with marginal growth, attached by rhizoidal filaments arising from lower cells and penetrating substrate; hairs absent.

Reproduction by quadriflagellate zoospores and biflagellate swarmers which may act as gametes.

The form of the prostrate system and the relation between this and the erect system needs clarification.

Until recently, two species, *Pseudopringsheimia confluens* and *P. fucicola* have been included in the genus and these are retained here. Morphologically they are very similar as they are found in the field, but they appear to differ in their life histories. The life history of *Pseudopringsheimia confluens* has been worked out completely in culture (Perrot, 1969); the culture work of Nielsen (1980) suggests that the life history of *P. fucicola* differs from that of *P. confluens*. She thinks that the morphology and mode of fragmentation of *P. fucicola* in culture resemble those of *Pseudendoclonium* and has transferred the taxon to *Pseudendoclonium* as *Pseudendoclonium fucicola* (Rosenvinge) Nielsen. The relationship of the two species of *Pseudopringsheimia* to one another and to *Pseudendoclonium* probably needs further clarification.

KEY TO SPECIES

1 Gelatinous cushions up to 75 μm thick, cushions distinct, cells 5–7 μm
    diameter, plants epiphytic on species of *Fucus* .... *P. fucicola*
2 Non-gelatinous cushions up to 250 μm thick, cushions becoming confluent,
    cells 10–12 μm diameter, epiphytic on stipes of *Laminaria* species .....
    .... *P. confluens*

Pseudopringsheimia confluens (Rosenvinge) Wille (1909b), p. 89.

Lectotype: C. Godthaab, West Greenland.

Ulvella confluens Rosenvinge (1893), p. 924.
Pseudopringsheimia penetrans Kylin (1910), p. 6.

Gametophyte forming firm, more or less confluent cushions on surface of host plant, consisting of closely appressed erect filaments with rhizoidal prostrate system penetrating between cells of host; cushions up to 260 μm thick in centre; erect filaments sparsely branched, of rows of 3–7, occasionally more, cells; cells rounded at apex of filament; chloroplast plate- or cup-shaped, parietal, with one pyrenoid; sporophyte, known only in culture, filamentous.

Reproduction by biflagellate isogametes, formed in gametangia terminal on erect filaments, or from cells deeper in cushion, $10–12 \times 27–44$ μm containing 30–40 gametes each $3–4 \times 2.5–3$ μm; also by quadriflagellate zoospores formed in terminal cells of sporophytic filaments, 6–12 per cell.

Chromosome no.: n = 6, 2n = about 12 Perrot (1969).

Found growing on stipes of species of *Laminaria* in the sublittoral region.

There are apparently only two records of this species in the British Isles, but it may have been overlooked: Dorset, Co. Dublin.

There are occasional records for the European Atlantic coast from Faroes, Norway, Sweden, France, Mediterranean.

On the Atlantic coast of North America, Newfoundland, Nova Scotia, Greenland.

There is no information on the seasonal behaviour of *Pseudopringsheimia confluens* in the British Isles. In France and Eastern Canada, the plants appear in the spring and remain through the summer months, they also reproduce during the whole of this period. The full seasonal picture is, however, not clear.

A life history involving a heteromorphic alternation between a cushion-shaped gametophyte and a filamentous sporophyte has been found in cultures by Perrot (1969). Meiosis has not been seen, but Perrot assumed that, from her chromosome accounts, it occurred in the sporangium.

It is not certain that *Pseudendoclonium confluens* is confined to *Laminaria* stipes. The present author has found plants with more or less confluent cushions and with the cell dimensions of this species on the stipes of *Fucus* species.

Pseudopringsheimia fucicola (Rosenvinge) Wille (1909b), p. 89.   Fig. 35

Holotype: C. Egedesminde, West Greenland.

Ulvella fucicola Rosenvinge (1893), p. 926.
Pseudendoclonium fucicola (Rosenvinge) Nielsen (1980), p. 135.

Plants forming discrete gelatinous cushions up to 1 mm in diameter, up to 75 μm thick, attached by rhizoids penetrating between cells of the host; erect system of rows of 3–6 (–7) cells, 5–7 μm diameter, 3–5 times as long as broad; chloroplast parietal in middle of cell; pyrenoid single.

Reproduction by quadriflagellate zoospores, 2–4 (–10) per sporangium, 4–5 × 5–6

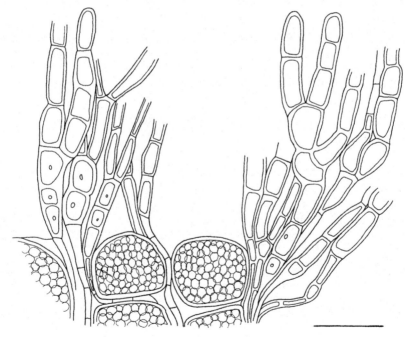

Fig. 35   *Pseudopringsheimia fucicola*
Plants epiphytic on *Fucus*. Note rhizoid penetration of host tissue (Turnaware Pt, Falmouth Estuary, Cornwall). Bar= 20 μm.

μm with large eyespot and posterior chloroplast, single pyrenoid; also by biflagellate swarmers, globular with eyespot, 3–3.5 × 3–4 μm; sporangia formed from uppermost cells of cushion.

Chromosome no.: not known.

Plants found growing as isolated cushions on *Fucus* species in the littoral and sublittoral regions.

Probably more widespread than is indicated by the records; Cornwall, Cheshire, Northumberland, Berwickshire, Shetland Isles, Co. Mayo.
Norway, Sweden, Denmark, Netherlands, Baltic Sea.
Nova Scotia, Greenland; Faroes, Iceland.

Plants are known to occur from February until June, but no information is available concerning the reproductive period in the British Isles. Swarmers are reported as forming in April in Denmark, in spring and summer in Sweden, in June in Greenland and from February to June in eastern Canada, but the full seasonal behaviour is not clear.
The life history is not known.

From her experimental work with this species, Nielsen (1980) believes that a heteromorphic alternation of generations, such as has been shown for *Pseudopringsheimia confluens*, is unlikely to occur in this species. She has shown that morphologically similar plants produce biflagellate swarmers and quadriflagellate zoospores. On the basis of this fact and also the fact that the morphology and mode of fragmentation of the plants resemble *Pseudendoclonium*, she transferred the taxon to *Pseudendoclonium* as *P. fucicola* (Rosenvinge) Nielsen.

ULVELLA Crouan frat.

ULVELLA Crouan frat. (1859), p. 288.

Type species: *Ulvella lens* Crouan frat. (1859), p. 288.

Pseudulvella Wille (1909), p. 90.

Thallus a prostrate disc formed of radiating dichotomously branched filaments coalescing laterally, at first monostromatic, later polystromatic in centre, attached by whole under surface of disc; cells uninucleate, usually without hairs, but forming long hyaline hairs under low nutrient conditions; chloroplast parietal, one pyrenoid present, or indistinct and apparently absent.
Reproduction by quadriflagellate zoospores formed in enlarged upper central cells of disc; biflagellate swarmers also reported.
Ulvella Crouan frat. is difficult to separate from the genera *Ochlochaete* Thwaites ex

---

Fig. 36   *Ulvella lens*
A. Disc epiphytic on *Plocamium*. (August 1971) B. Portions of discs on *Zostera angustifolia*. Note reproductive sporangia. (Fleet, Dorset; June 1976). Bar = 20 μm.

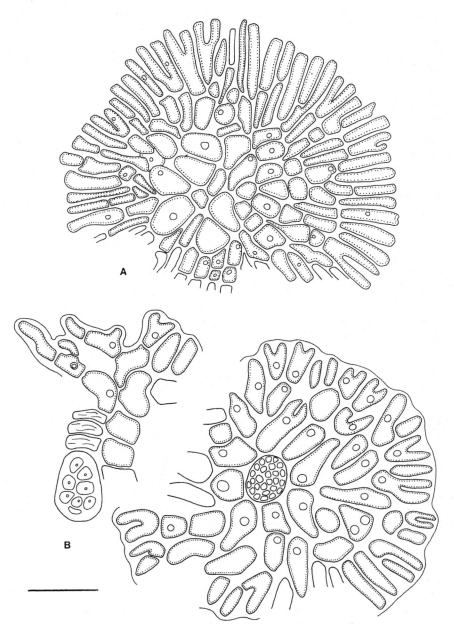

Harvey, *Chaetobolus* Rosenvinge and *Prinsheimiella* Höhnel (see discussion under *Ochlochaete* p. 116). *Pseudulvella* Wille and *Ulvella* Crouan frat. are here accepted as congeneric. *Pseudulvella* was separated from *Ulvella* by Wille (1909) on the basis of three characters: the cells were uninucleate, the chromatophore contained a pyrenoid and the zoospores had 4 flagella. These particular features were not referred to in the original description of *Ulvella*. Lami (1935) examined the type material of *Ulvella lens* in the Thuret Herbarium in PC and found that the cells were uninucleate and that most of the chromatophores had a distinct pyrenoid, though in some of the marginal cells they were very small and indistinct.

The idea that the cells of *Ulvella* might be multinucleate was introduced by Wille (1909) when he included *Dermatophyton* in the genus. This latter genus has now been removed (Ducker, 1958; Papenfuss, 1962), leaving *Ulvella* as a genus with only uninucleate cells. The only species included in *Pseudulvella* by Wille (1909) was *P. americana* (Snow) Wille (= *Ulvella americana* Snow), a freshwater alga. A marine species of *Pseudulvella*, *P. applanata*, was described by Setchell & Gardner in 1920. Nielsen (1977) finally showed that *Ulvella lens* has quadriflagellate zoospores, though it may also have biflagellate swarmers. *Pseudulvella applanata* was reported by Anand (1937) on chalk cliffs in Kent. There seems, however, no reason now, following the work of Nielsen (1977) to separate this from *U. lens*, or to maintain the genus *Pseudulvella* unless, as has been pointed out by Nielsen (loc. cit) *Pseudulvella americana* (Snow) Wille (= *Ulvella americana* Snow), the type species of the genus *Pseudulvella*, can be shown to have important characters other than those discussed above, distinguishing it from *Ulvella*.

Only one species accepted for the British Isles.

Ulvella lens Crouan frat (1859), p. 288.   Fig. 36

Holotype: PC. (Herb. Thuret). Rade de Brest on fragments of porcelain, shells and on other algae.

Pseudulvella applanta Setchell & Gardner (1920) p. 29.

Thallus of branched filaments radiating from a centre adhering to form a regular or somewhat irregular disc 1–2 mm in diameter, polystromatic in centre when mature, monostromatic towards periphery, attached to the substrate by the whole lower surface; central cells more or less rounded, (5–) 10–15 (–20) μm diameter; peripheral cells rectangular, 5–10 μm diameter, 2–4 times as long as broad, in radiating rows with simple or forked end cells; cells usually without hairs, but sometimes with long hyaline hairs; chloroplast with 1, occasionally 2 pyrenoids, pyrenoids sometimes indistinct or absent.

Reproduction by quadriflagellate zoospores formed in slightly enlarged upper central cells, 4–16 (–32) per sporangium; zoospores escape through hole in top of sporangium; biflagellate swarmers also reported.

Chromosome no.: not known.

Plants are found growing on pieces of pottery, shells, stones and epiphytically on other algae in the lower littoral and sublittoral regions to a depth of 20 m or more; also on *Zostera* leaves.

Probably much more frequent than is indicated by the published records.
Dorset, Devon, Kent.
Sweden, Netherlands, France.
Atlantic coast of North America from North Carolina to the Gulf of Mexico; Japan.

Plants can be found at almost any time of the year and may also be fertile in any
month of the year, but records are very incomplete. Sporangia have been found in
May and through the summer in Dorset. The life history is incompletely known. A
succession of asexual generations involving quadriflagellate zoospore production was
found in culture for this species by Nielsen (1977). Whether or not the plants can
produce biflagellate swarmers, as recorded by Wille (1909) and Kylin (1949), and, if
so, the part they play in the life history is not known.

The cells of *Ulvella lens* vary in size in different plants and also in the length and
shape of the peripheral cells, in the regularity of the disc, in its compactness and in the
degree to which the centre of the disc becomes polystromatic. Irregularities in the
compactness of the thalli can be produced by the substrate on which the plants are
growing (Nielsen, 1977). *Ulvella setchellii* Dangeard (1931) with rather longer peripheral
cells than *U. lens*, is said by Nielsen (loc. cit.) to maintain its differences under
identical condition in culture. She therefore maintains it as a separate species. There is
no definite record of *U. setchellii* in the British Isles.

There are very obvious similarities between the morphologies of *Ulvella lens* and the
described species of *Ochlochaete* (see discussion on p. 116).

PILINIA Kützing

Pilinia Kützing (1843), p. 273.

Type species: *P. rimosa* Kützing (1843), p. 273.

Acroblaste Reinsch (1879), p. 364.

Plant with both prostrate and erect systems; prostrate system of branched filaments,
sometimes compacting laterally to form dense covering on substrate; erect system of
simple or branched filaments, sometimes ending in long hyaline or multicellular hairs;
chloroplast parietal extending over the whole cell length, with or without pyrenoids.
  Reproduction by zoospores formed in terminal or lateral sporangia; where known,
zoospores biflagellate.
  The genus description given above includes a species, *P. earliae* recently described
from Florida, which has 3–8 (–12) pyrenoids in the chloroplast (Gallagher & Humm,
1980).
A number of species of *Pilinia* have been described for North Atlantic shores, but
their separation presents difficulties. The whole genus needs further investigation.

Only one species is recorded for the British Isles.

Pilinia rimosa Kützing (1843), p. 273.

Lectotype: L (938. 112. 621) Cuxhaven, Germany.

Plants forming a dense, spongy, mucilaginous crust on substrate, yellowish to olive green, consisting of a prostrate system of branched filaments, giving rise to erect unbranched or branched filaments up to 600 μm (–2 mm) high with cylindrical or slightly barrel-shaped cells, 5–10 (–19) μm diameter, 1–2 times as long as broad; chloroplast parietal, pyrenoids lacking or up to 2 per cell; filaments sometimes ending in hyaline hairs.

Reproduction by zoospores formed in enlarged apical cells of erect filaments, or sessile on basal system.

Chromosome no.: not known.

Plants grow on stones, shells and wooden structures, often associated with blue-green algae, sometimes epiphytic on larger algae, in the mid- to upper littoral region, but may occasionally occur in the lower littoral.

The species may be more common than is indicated in the records; Channel Islands, Dorset, Isle of Man, Berwickshire, Norfolk.

Sweden, Germany, France.

Atlantic coast of North America: Maine, Massachusetts, Florida.

Little is known about the seasonal behaviour of this species. Its occurrence has been described as sporadic in the Isle of Man where sporangia have been found in April (Knight, unpublished). The nature of the zoospores in the British Isles is not known, though they have been described as biflagellate on the American coast. The life history is not known.

The species *Pilinia rimosa* presents a number of problems, as has been pointed out by Papenfuss (1962) who discussed the species and the genus. The original description by Kützing (1843) made no mention of hairs and the fact that they were shown in the illustration of Newton (1931), might mean that her plant represented a different taxon. The chloroplasts of *P. rimosa* are usually depicted without pyrenoids and it has generally been assumed that they lack them. However, Humm & Taylor (1961) described two pyrenoids per cell for this species from Florida. They also illustrated hyaline hairs terminating some of the erect filaments similar to those of Newton's (1931) illustration. There is also uncertainty as to whether sporangia can occur on the prostrate filaments in *P. rimosa* or whether there is confusion with another taxon. The whole genus needs further investigation.

TELLAMIA Batters

Tellamia Batters (1895), p. 169.

Type species: *Tellamia contorta* Batters (1895), p. 169.

Thallus endozoic, of radiating branched cellular filaments; cells longer than broad, usually constricted at joints; filaments penetrating host, infection arising from a single cell or group of nearly isodiametric cells on host surface.

Reproduction by zoospores formed in swollen thallus cells which may become nearly globular.

Two species of *Tellamia* were described by Batters (1895), but later workers (Printz (1926), Levring (1937), Kylin (1949), Sundene (1953)) have failed to separate them.

The two species are described separately here on the basis of a reassessment of their delimitation (Mackie & Wilkinson, unpublished), but the distribution records are given for the genus. It seems that the failure of earlier workers to separate *T. intricata* and *T. contorta* may have resulted from the overlapping descriptions and poor figures presented by Batters. As a result of the recent reassessment based on examination of field material in shells of different ages, it is suggested that Batters probably included the older plants of *T. intricata* in his description of *T. contorta*, thus giving rise to confusion. Batters used the formation of ball-shaped clusters of cells in *T. contorta* as an important feature, but these appear to be so uncommon that it has not been possible to determine whether they are particularly associated with both or only one of the defined species and they have therefore been omitted from the present descriptions.*

KEY TO SPECIES

1 Both surface and penetrating cell systems present; filaments often coiling and twisting; cells more or less isodiametric; chloroplast lacking pyrenoid .
. . . . *T. contorta*

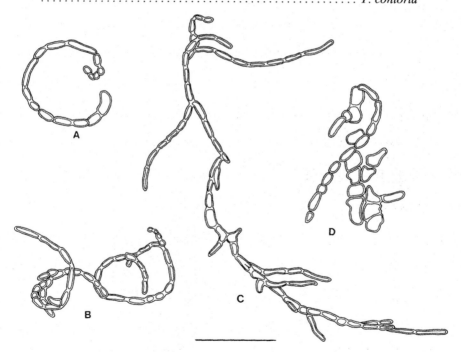

Fig. 37   *Tellamia contorta*. A–D. Simple and branched endozoic filaments. Bar= 25 μm.

* Genus currently considered to be monospecific, see Nielsen & McLachlan (1986. *Br. phycol. J.* **21**: 281–286) eds.

Only penetrating system present; cells oblong or spindle-shaped; chloroplast with 1 pyrenoid.................................... *T. intricata*

Tellamia contorta Batters (1895), p. 196.　　　　　　　　　　Fig. 37

Holotype: BM (Batters, 1895, mentioned that he had distributed specimens under a manuscript name to his friends). Weymouth.

Thallus consisting of both surface and penetrating systems; surface system of branched filaments; cells angular, polygonal or rounded sometimes isodiametric, 5 (3–9) × 4 (2.5–7) μm; penetrating system of filaments arising from surface system, penetrating at right angles to host periostracum surface, filaments often coiling and spiralling; branching relatively infrequent; secondary surface system may form where perforating filaments pass through shell surface; perforating cells 6 (2–10) × 2.5 (2–4) μm; chloroplast parietal, relatively imperforate; no pyrenoids visible.

Life history and reproduction largely unknown in this country.

Chromosome no.: not known.

Tellamia intricata Batters (1895), p. 196.　　　　　　　　　　Fig. 38

Holotype: BM (Batters, 1895 mentioned that he had distributed specimens under a manuscript name to his friends). Weymouth.

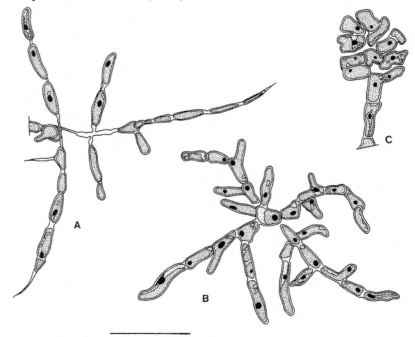

Fig. 38　*Tellamia intricata.* A–C. Portions of branched endozoic filaments. Bar= 25 μm.

Entocladia intricata (Batters) Kylin (1938), p. 70.

Thallus consisting of only branched filaments penetrating near surface of host, infection originating from a solitary surface cell: filaments much branched and interwoven; cells oblong or spindle-shaped, oval in old worn shells, sometimes isodiametric, in young cells 10 (4–23) × 4 (2.5–7) µm, in older cells 6 (2–10) × 3.5 (2–5) µm, average breadth of cells in filaments following cracks in eroded periostracum 3 µm; chloroplast parietal, perforated; pyrenoid single.

Reproduction and life history largely unknown in this country.

Chromosome no.: not known.

The genus occurs in the periostracum of species of *Littorina*, mainly *L. littoralis* in the mid- and lower littoral region where the host shell is found creeping on larger red and brown algae. It is found predominantly in living shells of *Littorina*; in dead and very old living shells the periostracum becomes eroded away so that the habitat for *Tellamia* is lost.

Tellamia has so far been found on all shores examined, where the host shell was present from the Orkney Islands to the south coast of England, also in Co. Wexford, Co. Galway and Co. Cork. It is likely therefore that it is common all round the British Isles.

Norway, Sweden, Denmark, France (Brittany).

Atlantic coast of North America in Maine and southern Massachusetts.

Plants are known to be present from March until August; possibly they are present throughout the year. So far sporangia have been recorded only in spring (Parke, 1935).

Tellamia is the only genus known to perforate the periostracum of *L. littoralis* in Britain. Therefore its presence can easily be confirmed if other surface species of algae are removed by brushing the shell and loosening the periostracum from the calcareous part of the shell. The periostracum can then be laid flat on a microscope slide but it must be noted that brushing the shell may remove some surface cells of *T. contorta*, so that it is necessary also to examine some unbrushed shells.

CLADOPHORALES West

CLADOPHORALES West (1904), p. 56.

Thallus of unbranched or branched filaments, unattached or attached by disc or rhizoids; no distinct prostrate system; except during reproduction apical cells inconspicuous, narrower than cells lower down filaments; thallus branches little differentiated with only occasional rhizoidal branches; lateral branches arising at an angle to main axis, (usually just below a cross wall), becoming pseudodichotomous by evection; cell division both apical and intercalary: cells semicoenocytic, multi-nucleate, no synchronisation of nuclear and cell division; new cross walls formed by ingrowth of annular rim of thickening on inner parent cell wall; chloroplast a mesoplastid made up of anastomising, but originally separate plates; pyrenoids numerous.

Reproduction by biflagellate gametes and quadriflagellate zoospores.

The Cladophorales form a much more coherent order now that the Acrosiphoniaceae has been removed to the Ulotrichales with which it has much more in common. The Acrosiphoniaceae differ from the Cladophorales in that in this order a definite prostrate system is lacking, the chloroplast is a 'mesoplast' made up of anastomosing, but originally separate, plates, cell division and nuclear division are not synchronised, the life history is an isomorphic alternation of generations or a modification of this, crystalline cellulose I is present in the cell walls (Nicolai & Preston, 1952, 1959) and the breakdown products of the polysaccharides are mainly arabinose, galactose and xylose (O'Donell & Percival, 1959): see also discussion under Acrosiphoniaceae p. 70.

One family, Cladophoraceae.

CLADOPHORACEAE Wille

CLADOPHORACEAE Wille in Warming (1884), p. 30.

Pithophoraceae Wittrock (1877), p. 47.

Diagnosis of the family is the same as for the order.

Four genera occur in the British Isles: *Chaetomorpha*, *Cladophora*, *Rhizoclonium* and *Wittrockiella*.

With the removal of the genera *Urospora* and *Spongomorpha* to families within the Ulotrichales, the Cladophoraceae forms a much more coherent family. The Cladophoraceae includes *Cladophora* and *Wittrockiella* which are branched, and *Chaetomorpha* and *Rhizoclonium* which are unbranched. Attempts to separate genera and species within the unbranched complex have often led to confusion. The extent to which the main characters used for separation are modified by the environment is still not too well known. These characters include colour, cell shape, cell dimensions, types of basal attachment and the presence or absence of rhizoids and short lateral branches.

Both Price (1967) and Nienhuis (1975) have shown rhizoid formation to be dependent on environmental conditions. Price (1967) worked out mean sizes and confidence limits for cells in samples of unbranched Cladophoraceae collected at different times of the year from selected sites and from different locations in the British Isles, from the south coast to Fair Isle and the Shetland Isles. She recognised 5 taxa of unbranched plants: *Rhizoclonium riparium* (Roth) Harv. (= *R. tortuosum* (Dillw.) Kütz.), *R. implexum* (Dillw.) Kütz., *Chaetomorpha capillaris* (Kütz.) Børg. (= *C. mediterranea* (Kütz.) Kütz.), *C. linum* (Müll.) Kütz., attached *C. aerea* (Dillw.) Kütz. (which she considered conspecific with the unattached *C. linum*) and *C. melagonium* (Weber & Mohr) Kütz., which could also exist attached or unattached. Nienhuis (1975) has shown that the characters used to distinguish the two species of *Rhizoclonium* are unstable and that the ranges in cell dimensions for *R. riparium* include those of *R. implexum*. The data given by Price (1976) could also be interpreted in this way. Taking the above works into account, two genera, *Rhizoclonium* and *Chaetomorpha*, the former with a single species, *R. tortuosa*, and the latter with three species, *C. linum*, *C. mediterranea* and *C. melagonium*, are recognised here. *R. lubricum* was placed in a new genus *Lola* by A. & G. Hamel (1929) on the grounds that it differed from *Rhizoclonium* in having more than 4 nuclei per cell (*Lola* 5–16) and from *Chaetomorpha* by its heterogamous reproduction. The plants are intermediate between *Rhizoclonium* and *Chaetomorpha*, particularly in mean cell size variation pattern. *Lola* is not accepted here, but see Abbott & Hollenberg (1976).

CHAETOMORPHA Kützing nom. cons.

CHAETOMORPHA Kützing (1845), p. 203.

Type species: *Chaetomorpha melagonium* (Weber & Mohr) Kützing (1845), p. 204. See Silva (1950), p. 254.

Chloronitrum Gaillon (1828), p. 389.
Spongopsis Kützing (1843), p. 261.
Lola A. & G. Hamel (1929), p. 1094.

Plants solitary or tufted, of uniseriate unbranched filaments attached by lobed or discoid base or by well developed basal cell, or an entangled mass of filaments with crisp texture; diameter of cells of attached filaments increasing from base to apex of filament; cells of unattached filaments of equal diameter throughout; cells cylindrical without constrictions or with some degree of constriction between cells or cells barrel-shaped to globose; cell walls striate; cells multinucleate; chloroplast parietal, reticulate with numerous pyrenoids.

Reproduction by quadriflagellate zoospores formed in undifferentiated cells, released through lateral pore; biflagellate gametes formed in undifferentiated cells; gametes isogamous; plants dioecious; vegetative reproduction by fragmentation; akinetes recorded.

Life history, where known, appears to be an isomorphic alternation of gametophyte and sporophyte generations.

KEY TO SPECIES

1   Filaments basally attached, cell width increasing from base to apex, well
     developed basal cell   2
    Filaments unattached, cells of constant width along the filament   3
2   Maximum cell width (70–) 210–1050 µm, cells 2–3 times as long as broad,
     basal cell as long as 3 mm, adjacent cells 8–14 times as long as broad,
     septa constricted giving plant a beaded appearance, colour dark blue-
     green with a glaucous sheen . . . . *C. melagonium*
    Maximum cell width (28–) 128–585 µm, cells 1–2 times as long as broad,
     basal cell approx. 1 mm long consisting of 1–9 fused cells, adjacent cells
     1–4 times as long as broad, septa slightly constricted, plant texture
     rough, colour dark to yellow green . . . . *C. linum*
3   Cell width range (30–) 40–80 (–100) µm, filaments forming tangled skeins
     on rocks and in pools, often associated with *Corallina officinalis* L.,
     very occasionally to be found attached by a non-differentiated basal
     cell, colour bright green, yellowing in summer . . . . *C. mediterranea*
    Cell width range 200–300 µm forming thick entangled masses in brackish
     ditches or mud banks, colour light to dark green, variable, cells only
     slightly constricted at septa . . . . *C. linum*

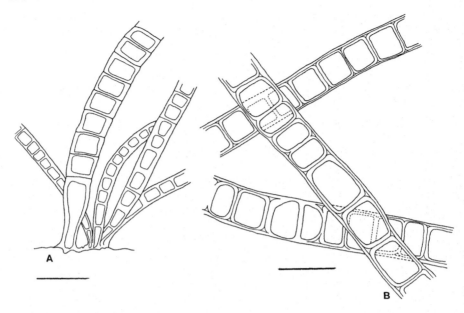

Fig. 39   *Chaetomorpha linum*
          A. Bases of filaments with elongate rhizoidal cells. (Orkney). Bar= 500 µm.
          B. Portions of erect filaments. (Orkney). Bar= 200 µm.

Chaetomorpha linum (O. F. Müller) Kützing (1849), p. 379.   Fig. 39

Type: C? Nakskov Fjord and Rødby Fjord, Denmark. The latter is now reclaimed
land. Abbott & Hollenberg (1976) say that the type material is from England and
Wales, without explanation.

Conferva linum O. F. Müller (1778), t. 771, f. 2.
Chaetomorpha aerea (Dillwyn) Kützing (1849), p. 379.
Conferva aerea Dillwyn (1806), p. 46.
Conferva capillaris sensu Dillwyn (1805), p. 46 non Linnaeus (1753), p. 1166.

Filaments unattached or basally attached; when unattached, cells 200–300 µm broad,
1–2 times as broad as long, cell width constant along filament; when attached, width
gradually increasing from base to apex of filament; basal attachment rhizoidal or by
means of a membranous disc; range of maximum cell width (28–) 120–535 µm, mean
maximum cell width 126–400 µm, cells immediately above basal cell 1–2 times as long
as broad, not noticeably constricted at septum, constriction more pronounced towards
distal end where cells ½–1½ times as long as broad; colour variable, often dark green
at base becoming lighter green towards distal end; texture firm.

Reproduction by quadriflagellate zoospores, 12 × 7 µm, formed in sporangia only slightly modified from vegetative cells; also by biflagellate gametes, 11 × 5 µm; plants dioecious.

Chromosome no.: for plants identified as *Chaetomorpha linum*
                   attached form 2n = 36   Patel, (1961)
                              2n = 18   Sinha (1958)
               unattached form 2n = 36   Sinha (1958), Patel (1961)
           for plants identified as *Chaetomorpha aerea*
                              2n = 18   Sinha (1958)
                              2n = 12   Patel (1961)
                              2n = 20, n = 10 Hartmann (1929)

Unattached plants form entangled masses in seawater, ditches or on mud banks; attached plants occur in shallow, often sandy pools in the mid- and lower littoral region and in the shallow sublittoral.

Both unattached and attached forms occur all round the coast of the British Isles.
North European Atlantic coast from Scandinavia and the Baltic Sea south to the Mediterranean, including Morocco.
North American Atlantic coast from Labrador south to Florida; Bermuda; California; Mauritius; Sri Lanka.

Both attached and unattached plants are abundant in spring and summer, but may be found at any time of the year. For attached plants, gametes have been recorded mainly from April to August and zoospores mainly from August until November (Price, 1967), but fertile plants can be found all the year round (Knight & Parke, 1931). For the attached form an isomorphic alternation of generations has been reported by Hartmann (1929): the data give by Price (1967) would support this. The unattached form has not been found in a fertile condition in the field, but Kornmann (1972) obtained both zoospores and gametes in culture from material collected from the type locality in October.
Maximum cell diameter varies with environmental conditions on both a seasonal and a geographical basis and plants growing under estuarine conditions tend to be narrower than those found on the open shore (Price, 1967).
It is still not generally agreed that the unattached *Chaetomorpha linum* and the attached *C. aerea* are conspecific. Some authors agree with Taylor (1937) that until intermediate forms are found, the taxa should remain distinct. Valet (1961) separated the two species on morphological grounds: cells of *C. linum* cylindrical with extreme diameters of 300–350 µm and those of *C. aerea* globose with extreme diameters of 500–750 µm. Price (1967) disagreed since, in her work, intermediate sized cells of both types could be found, particularly during reproduction. In sheltered Danish waters, Christensen (1957a) found that unattached plants can originate from attached plants characteristic of *C. aerea*. This was supported by Price who found that attached material, collected from the field and cultured, lost its straight form and became curled and matted. Kornmann (1969, 1972) on the other hand, believed the two species to be distinct. The appearance of the filaments differed, the attached having distinct constrictions at the cross walls while the unattached had no such constrictions though the diameter of the cell was slightly reduced in the middle. He also recorded that the attached form has an alternation of isomorphic sporophyte and gametophyte phases in

its life history, while the free-floating form has never been found fertile in the field. The chromosome counts of Sinha (1958) and Patel (1961) might suggest that the two taxa are distinct. However, the actual numbers do not agree.

Bearing in mind all of the evidence, it has been decided for the purpose of this flora, to regard *C. linum* and *C. aerea* as conspecific.

Examination of material referred to *C. crassa* (Ag.) Kütz. by Price (1967) indicates that this species may be synonymous with *C. linum*. However, until material in Agardh's herbarium has been seen, the question of synonymy remains uncertain.

Chaetomorpha mediterranea (Kützing) Kützing (1849), p. 381.   Fig. 40

Holotype: L. Livorno, Italy.

Spongopsis mediterranea Kützing (1843), p. 261.
Rhizoclonium capillaris sensu Kützing (1847), p. 376 non *Conferva capillaris* Linnaeus (1753), p. 1166.
Chaetomorpha tortuosa sensu Kützing (1849), p. 376 non *Conferva tortuosa* Dillwyn (1809), p. 46.

Unbranched filaments entangled to form a spongy mass, basally attached only in early stages of development; lateral rhizoids absent; texture firm and crisp; colour variable, grass green to yellow green; cells (30–) 40–80 (–100) μm broad, (1) 2–3 times as long as

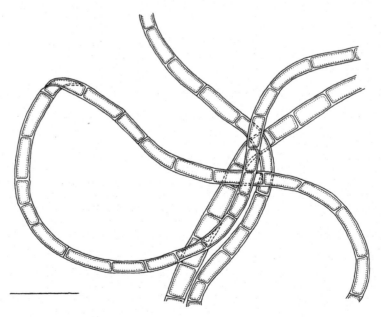

Fig. 40   *Chaetomorpha mediterranea*
Portions of entangled erect filaments. (Orkney). Bar= 200 μm.

broad; 20–50 nuclei per cell; chloroplast parietal, reticulate; cell walls 3–4 μm thick, striated and lamellate, with slight constriction at septum.

Reproduction mainly vegetative; also by quadriflagellate zoospores and biflagellate isogamous gametes.

Chromosome no.: n = 11 (Kornmann, 1972)
n = 11, 2n = 22 (Patel, 1961 for *Rhizoclonium tortuosa* Kütz. ?)

Found in tangled masses on rock and over other algae growing on rock and in pools in the lower littoral region, often associated with *Corallina officinalis*; it is also found in the shallow sublittoral. Young filaments may be attached. Plants can be found in estuaries and under brackish water conditions.

Distributed round the whole coast of the British Isles and generally common.

European Atlantic coast from North Norway south to Portugal: also in the Mediterranean; not found in Troms nor in Finnmark (N. Norway) by Jaasund (1965) in his collections made there between 1949–1960 though reported there earlier by Foslie (1890) and Kjellman (1883).

Greenland; Iceland, Faroes.

Plants are present all the year round, abundant in summer, less so in winter. Swarmers have been found in July and young plants in September, but probably vegetative reproduction is dominant in the field (Price, 1967).

The life history has not been worked out for plants in Britain, but in Helgoland, Kornmann (1972 as *C. tortuosa*) found an isomorphic alternation of gametophyte and sporophyte generations: the plants were dioecious.

Mean cell measurement for populations vary with variations in environmental conditions within the limits given for the species (Price, 1967).

Chaetomorpha melagonium (Weber & Mohr) Kützing (1845), p. 204.

Lectotype: original illustration (Weber & Mohr 1804, pl. 3, figs 2a, 2b) in absence of material. Varberg, Sweden.

Conferva melagonium Weber & Mohr (1804), p. 194.
Chaetomorpha atrovirens Taylor (1937b), p. 227.

Taxonomic complications arise due to the existence of both attached and unattached forms of *Chaetomorpha melagonium* (Web. & Mohr) Kütz. Foslie (1890) described two forms, the unattached forma *typica* Kjellm. and the attached forma *rupincola* Areschoug. Collins (1905) and Farlow (1881) suggested that forma *typica* was based on the species *Conferva piquotiana* Mont. (1845) which is apparently a later stage of the attached form. Taylor (1937) was aware that a number of authors accepted the unattached form as a later developmental stage of the attached form, but, in the absence of adequate supporting evidence, he was not convinced. Material collected by Burrows in Canada in 1960 (now in Liverpool Museum Herbarium), identified as *Chaetomorpha atrovirens* by Taylor, was accepted by Price (1967) as an unattached form of *C. melagonium*. In the absence of conclusive evidence of the distinctness of the attached and apparently unattached forms, they have both been included in the description of *C. melagonium* (Web. & Mohr) Kütz. given here.

Filaments simple and unbranched, normally basally attached, stiff, straight, gregarious or solitary, up to 60 cm long; colour dark blue-green with distinct glaucous sheen, texture coarse; filaments attenuate at base, basal cell up to 3 mm long representing (1) 2–4 coalesced cells; attachment by rhizoidal system; cells immediately above basal cell, 8–14 times as long as broad, cylindrical, no marked constriction at septum; distal cells barrel-shaped or sub-globose, with marked constriction at septum, giving plants a beaded appearance; cells 2–4 times as long as broad, cell width (70–) 210–1050 µm; cell wall approximately 8 µm thick, distinctly lamellate, cellulose microfibrils orientated in various planes (TEM); chloroplast parietal, reticulate, pyrenoids numerous, nuclei at least 100 per cell; meristem in lower part of filament, diffuse, distal cells of increasing maturity.

Reproduction by biflagellate gametes and quadriflagellate zoospores; gametangia and zoosporangia only slightly modified vegetative cells, produced from all cells except the few at the filament base; gametes isogamous with size ca. 15 × 7 µm, zoospores 11 × 8 µm; plants monoecious or dioecious. According to Kornmann (1972) the plants produce asexual biflagellate swarmers mixed with a few quadriflagellate swarmers. Vegetative reproduction by budding of upright filaments from rhizoidal base, thus replicating the haploid or diploid phase.

Chromosome no.: n = 18 Sinha (1958).
n = 12, 2n = 12, 24 Patel (1961).

Plants are found on sublittoral rocks and in lower shore pools.

Widespread round the whole coast of the British Isles though it rarely occurs in great quantity.

North European coast from Norway to the Atlantic coast of France; Spitzbergen, Murman Sea, White Sea.

Iceland, Faroes, Greenland, western Atlantic coast from the Canadian Arctic south to Cape Cod; Collins (1908) gives Alaska, but this is not recorded by Scagel (1966).

Plants are inconspicuous over the summer, more abundant in autumn and winter.

Swarmers are produced in spring associated with equinoctial spring tides, all but the basal one or two cells emptying completely, leaving 'ghost' filaments. The life history has not been completely worked out but circumstantial evidence suggests that it is an isomorphic alternation of gametophyte and sporophyte generations. Each phase appears to last for about 12 months. Köhler (1956) recorded that only a minute proportion of the swarmers produced are actually viable.

Maximum mean cell size is achieved in winter prior to reproduction and is lowest in summer when young plants are appearing.

CLADOPHORA Kützing, nom. cons.

CLADOPHORA Kützing (1843a), p. 262.

Type species: *Cladophora oligoclona* (Kützing) Kützing (1843a), p. 262.

Prolifera Vaucher (1803), p. 118.
Polysperma Vaucher (1803), p. 90.

Annulina Link (1820), p. 4.
Aegagropila Kützing (1843), p. 272.
Basicladia Hoffman & Tilden (1930), p. 380.
Rama Chapman (1952), p. 55.

Thallus a uniseriate, usually highly branched filament of cells; cells decreasing in size from base to apex; filaments attached by a rhizoidal system or a discoid holdfast, or unattached, forming entangled or ball-shaped masses; growth of filaments by apical, often also by intercalary division; branches inserted laterally or apically on cells, becoming pseudodichotomous by evection; cells cylindrical, barrel-shaped or club-shaped, multinucleate; chloroplast parietal, reticulate with many pyrenoids; cell wall sometimes thick and striated.

Reproduction usually by quadriflagellate zoospores and biflagellate gametes, also vegetatively by fragmentation and by formation of thick-walled akinetes.

Life history, where known, an isomorphic alternation of generations.

Reference should be made to the work of van den Hoek (1963), the most comprehensive work on the European species of *Cladophora* and that on which the descriptions of the British *Cladophora* species have been made here. Account has also been taken of the work of Söderström (1963), Archer (1963) and other authors. The dimensions of cells and reproductive bodies are those given by van den Hoek (loc. cit.) except where other measurements fall outside these. For a complete synonymy of this genus see van den Hoek (loc. cit.).

KEY TO SPECIES
(adapted from van den Hoek, 1963)

1　Plants not more than 1.5 cm high, densely branched, branches apically
　　inserted; holdfast disc-shaped . . . . *C. pygmaea*
　　Plants more than 1.5 cm high; plants unattached or with rhizoidal
　　holdfast 2
2　Insertion of branches lateral with vertical or steeply inclined cross walls . . 3
　　Insertion of branches apical, with oblique cross walls becoming pseudo-
　　dichotomous by evection 6
3　Cells at base of plant giving off rhizoids with annular thickenings at basal
　　pole; basal cells club-shaped, long, up to 1 cm . . . . *C. prolifera*
　　No rhizoids with annular thickenings; basal cells long and somewhat club-
　　shaped; axes with no intercalary divisions except when forming
　　sporangia 4
　　No rhizoids with annular thickenings; basal cells not very long 5
4　Thallus unattached, tangled round bases of *Zostera* plants or other
　　algae . . . . *C. retroflexa*
　　Thallus attached by rhizoidal holdfast . . . . *C. pellucida*
5　Plants densely and irregularly branched; basal cells cylindrical or only
　　very slightly club-shaped; rhizoidal branches formed from any cell: cell
　　walls thin . . . . *C. coelothrix*
　　Plants densely and irregularly branched; basal cells club-shaped; rhizoidal
　　branches formed from any cell; cell walls thick and lamellate especially
　　in basal region . . . . *C. aegagropila*

6   Plants consisting of pseudodichotomously branched main axes lined with
    branches of different ages, younger branches intercalated between
    older ones; intercalary growth dominant; feebly acropetal organisation
    only in young plants and in plants exposed to wave action   7
    Plants consisting of long unbranched or sparsely branched axes growing
    mainly by intercalary division, at times with many almost simultaneous
    proliferations   9
    Plants consisting of pseudodichotomously branched main axes, ending
    above in acropetally organised, often falcate, branch systems with
    dominant apical growth, intercalary divisions beginning well below
    apex, increasing basipetally; in quiet water forms with intercalary
    growth may be dominant, but axes never lined with rows of branches of
    mixed ages   10
7   Plants dark green, lacking rhizoids and with strongly bent and crooked
    axes, tangled round *Zostera* plants or round other algae; forming
    aegagropiloid aggregates; branches curved, often sharply pointed . . . . .
    . . . . *C. battersia*
    Plants with rhizoids   8
8   Apical cells 90–195 μm diameter . . . . *C. hutchinsiae*
    Apical cells 40–80 μm diameter; plants stiff; branches straight or slightly
    curved outwards, up to 6 formed on a single axial cell . . . . *C. rupestris*
    Apical cells 18–24 μm diameter in pale green insolated plants, 50–70 μm in
    dark green, shaded plants; apical cells pointed; bases of branches
    decurrent with axial cells especially in lower part of plant . . . . *C. sericea*
    Diameter of apical cells 10–16 μm in pale green, insolated plants,
    32–41 μm in dark green, shaded plants, apical cells rounded at tip . . . . .
    . . . . *C. albida*
9   Diameter of axes 16–27 (–38) μm; diameter of apical cells 16–20 μm;
    plants attached by rhizoids or unattached . . . . *C. globulina*
    Diameter of axes 30–175 μm; diameter of apical cells 22–39 μm; plants
    unattached . . . . *C. liniformis*
10 Apical cells 90–160 μm diameter; branches mostly strictly acropetally
    organised, sometimes consisting of old axes lined with opposite
    proliferations . . . . *C. lehmanniana*
    Apical cells 35–45 μm diameter in pale green, insolated plants, 80–110 μm
    in grass green plants from shaded sites; apical cells cylindrical or very
    slightly tapering with rounded ends . . . . *C. laetevirens*
    Apical cells narrower   11
11 Apical cells 20–25 μm diameter in pale green, insolated plants to 55–90 μm
    in dark green plants from shaded sites; marine plants penetrating into
    brackish water with salinity as low as 7‰; plants attached or free
    floating . . . . *C. vagabunda*

Fig. 41   *Cladophora aegagropila*
          Portions of erect filaments. A, B. Branch apices. C. Mid region of plant. D. Lower
          region of plant with signs of loss of polarity. E. Irregularly branched filaments with
          young apices and swollen cells. Filaments derived from a floating ball and lack
          polarity. A–D. Orkney. E. Dec. 1977; H. Blackler. Bar= 200 μm.

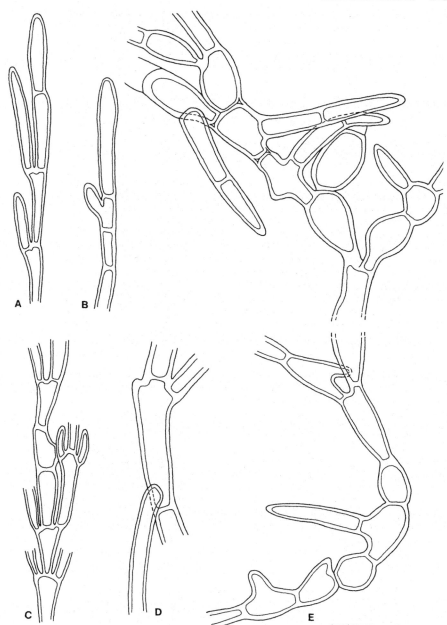

Apical cells 14–20 μm diameter in pale green, insolated plants to 30–35 μm in plants from shady sites; axes sometimes with irregular swellings; branches sometimes distinctly incurved; marine plants with very constant acropetal branch organisation persisting even in floating plants; plants attached or free floating .... *C. dalmatica*

Cladophora aegagropila (Linnaeus) Rabenhorst (1868), p. 343–344.   Fig. 41

Type: LINN (No. 1277.49; see van den Hoek, 1963, p. 51). Sweden.

Conferva aegagropila Linnaeus (1753), p. 1167.
Cladophora brownii (Dillwyn) Harvey (1846), pl. XXX.
Conferva brownii Dillwyn (1802–1809), p. 58.

Plants dark to bright green, aggregated into stiff tufts forming cushions or carpets 1–3 cm high, attached by rhizoids, or forming unattached matted balls up to 20 cm or more in diameter, highly and often irregularly branched; apical cells frequently lost; branches formed at upper ends of young axes, often slightly sub-terminal on cells, with vertical or steeply inclined cross walls, polarity later lost; rhizoids may form laterally from basal ends of cells, these attaching by coralloid holdfasts; rhizoids rare in loose balls; apical cells 30–70 μm diameter, cells of lateral branches 30–100 μm diameter, cells of main axes up to 200 μm diameter; cells of main axes club-shaped or irregularly swollen; cell contents, except in young branches, very dense; cell walls thick and striated up to 20 μm thick in main axes, especially at base of plant.

Reproduction mainly by thallus fragmentation, though biflagellate swarmers have been reported (Nishimura & Kanno, 1927).

Chromosome no.: n = 12, 2n = 24 Sinha (1958) (Van den Hoek, 1963 comments that the chromosone numbers suggest that meiosis precedes spore-formation).

The species occurs in the sublittoral region of freshwater lakes and brackish lagoons and in the littoral region of shores washed by water of low or fluctuating salinities; it is not found under fully marine conditions.

The species is not common in Britain, but where found it occurs more in freshwater than in brackish conditions.

The very short (1–3 cm) attached plants (= *C. brownii* (Dillwyn) Harvey) have been found recently in the littoral region, on a soft muddy/gravel substrate, in the Fleet, Dorset where the water salinity fluctuates from (5–) 15–30‰. This type of plant was reported from Cornwall and Co. Wicklow, Ireland, during the last century (Harvey 1846). Apart from a number of freshwater habitats, larger unattached plants have been found recently in the Orkney Islands and in south Uist, Western Isles.

The species as a whole is distributed in Europe from Finland to the Mediterranean, Switzerland, Austria; Faroes.

The two types of plant referred to above appear distinct in their morphology and in

---

Fig. 42   *Cladophora albida*
A, B. Portions of upper branched filaments. Apex in B is preparing to swarm. Bar= 200 μm. C. Portion of reproductive filament. (Orkney). Bar= 50 μm.

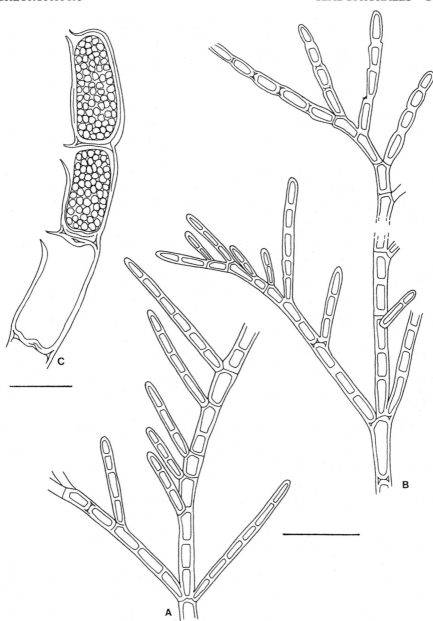

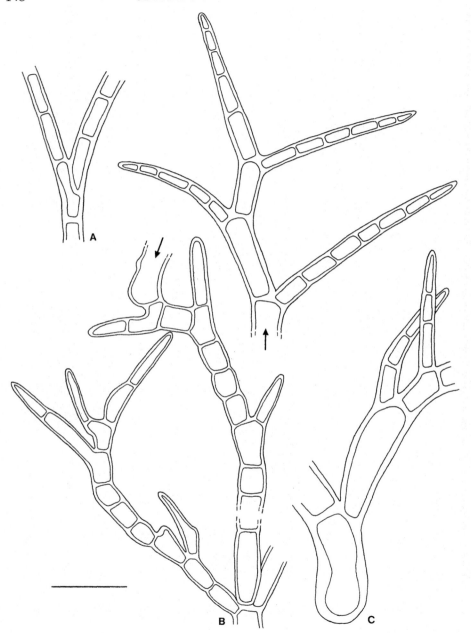

their distribution, but van den Hoek (1963) regards them as habitat variants; he refers to a wide range of morphological variation with no distributional discontinuities.

Swarmers of unknown type have been seen in plants from south Uist in late summer. The life history of the species is not known; spread appears to be mainly vegetative, by fragmentation.

In the unattached aggregates of plants, the apical cells are frequently lost and branching is diffuse. Polarity tends to be lost and branches form in all positions on the cells, frequently three or more branches per cell, some of them downwardly growing; the main axes have dense contents and are swollen and irregular in shape.

Unattached balls continue to grow for up to 16 years or more and may become more than 20 cm in diameter, frequently hollow in the centre (Lorentz, 1855).

Cladophora albida (Hudson) Kützing (1843), p. 276.   Fig. 42

Lectotype: OXF (*Conferva marina tomentosa, tenerior & albicans* in herb. Dillenius). Probably from Selsey, see van den Hoek, 1963; Scagel (1966) gives the type locality as 'Torquay, England', without explanation.

Conferva albida Hudson (1778), p. 595.

Plants 5–15 (–50) cm long, richly branched with a distinct spongy texture and very fine construction; whitish yellow to bright olive green in colour; filaments straight or curved becoming progressively narrower above with little constriction between cells except when reproducing; growth of axes mainly by intercalary divisions, the new cells giving rise to younger branches intercalated between older branches which line the axes, ultimate branches pectinate, occasionally opposite; branches cut off at apical poles of cells by oblique cross walls, later evection forming pseudodichotomies, with eventually 2–3 branches per axial cell, branches sometimes reflexed; cell walls clearly stratified; apical cells rounded at tip, 10–16 μm diameter in pale green plants, 32–41 (–50) μm diameter in dark green plants; main axes 32–90 μm diameter.

Reproduction by quadriflagellate zoospores and biflagellate gametes formed in slightly swollen terminal and subterminal cells of axes, following active intercalary divisions; biflagellate zoospores also sometimes formed; zoospores on average 13.8 × 6.5 μm, gametes 8.5 × 4.0 μm.

Chromosome no.: 2n = 24 Patel (1961).

The plants grow throughout the littoral region, in brackish pools at the highest levels, as undergrowth to larger algae lower down, on overhanging rocks and in rock crevices; they grow also in shallow lagoons and inlets with sandy or muddy bottoms where there may be considerable fluctuations in salinity and temperature; epilithic, epiphytic, or epizoic.

Common all round the coast of the British Isles.

Scandinavia south to the Mediterranean; Baltic Sea, Bosphorus, Black Sea.

Fig. 43   *Cladophora battersia*
A–C. Portions of erect filaments. Drawn from E. A. L. Batters material of *Cladophora corynarthra* var. *spinescens*. Bar= 200 μm.

North American Atlantic coast from Newfoundland south to Florida; reported also from North American Pacific coast from North Washington to El Salvador; Nigeria, Ghana, Cameroon; Japan.

Found on British shores mainly between April and September. The plants may overwinter in the form of persistent basal branch systems with thick-walled cells or as undeveloped germlings. Swarmers have been recorded in the summer months, though rarely, in this country where the life history has not been worked out, but in Sweden, Bliding (1936) found an isomorphic alternation of gametophyte and sporophyte generations.

The species is readily recognised by its small cell size and its mode of terminal branching. In exposed sites the plants are usually less than 5 cm high and may form extensive carpets; in sheltered sites they can reach 15 cm or more. Under very sheltered conditions the branches may become distant from one another so that the plants resemble the sheltered water form of *Cladophora sericea*. Under strong illumination the cells of *C. albida* are narrower and much paler green than under low illumination.

Two varieties of *C. albida*: *C. albida* var. *albida* and *C. albida* var. *biflagellata* are recognised by van den Hoek (1963). The two varieties have only minor morphological differences, but they do differ in their life histories, with the latter forming only biflagellate neutral swarmers.

Cladophora battersii van den Hoek (1963), p. 92.   Fig. 43

Lectotype: BM (see van den Hoek, 1963, p. 92). Roundstone, Co. Galway, Ireland.

Cladophora corynarthra var. *spinescens* Batters (1900), p. 370; non *Cladophora spinescens* Kützing (1849), p. 418.

Plants 1–3 cm high, rather stiff, usually unattached, forming dark green aegagropiloid masses; axes curved and bent; growth mainly intercalary; axes pseudodichotomously branched with younger branches intercalated between older branches; branches narrowing towards apices to give a spinous appearance; branch initials inserted apically on cells of axes, sometimes almost at right angles to axes; cell walls thick and striated; apical branches 40–75 μm diameter, main axes 90–115 μm diameter.

Reproduction vegetative by fragmentation; no asexual or sexual process known.

Chromosome no.: not known.

Plants occur tangled around bases of *Zostera* plants in quiet bays and lagoons; also on small stones and tangled around other algae in very sheltered sites in the lower littoral and shallow sublittoral.

This is a very rare *Cladophora* species. It was found at Portland, Dorset, by Holmes in 1895 and again by the present author in 1977. It was reported from Roundstone Bay, Co. Galway, by Batters (1902) and there are three records for the west of Ireland

Fig. 44   *Cladophora coelothrix*
A–E. Portions of erect filaments. Bar= 200 μm.

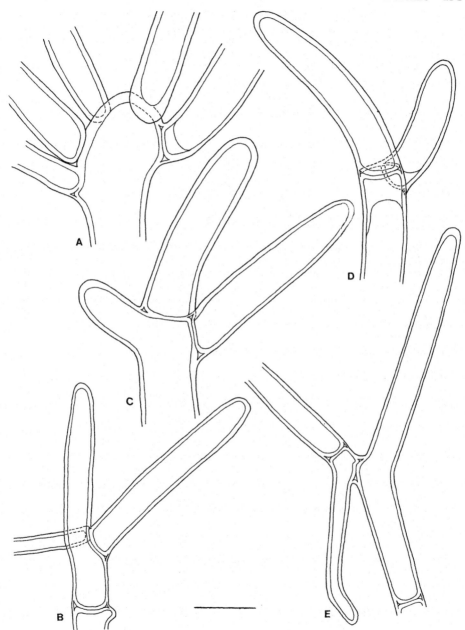

by Cotton (1912); Clare Island, Co. Mayo, Roundstone Bay, Co. Galway and Valencia Island, Co. Cork. Whether or not the species is still there is not known. There is a record from Guernsey, Channel Islands, based on a specimen by Lyle, 1914 (BM), (see van dan Hoek, 1963), but whether it still occurs there is not known.

The species was reported from France and Yugoslavia early this century, but there is only one recent record in France, that of Parriaud (1960), (see van den Hoek, 1963).

The species has not been found forming swarmers and the life history is unknown. The plants probably reproduce only by fragmentation and have spread little over the last hundred years.

Cladophora coelothrix Kützing (1843), p. 272.   Fig. 44

Holotype: L (No. 937. 278. 392, leg. Meneghini; see van den Hoek, 1963). Livorno, Italy.

Cladophora repens (J. Agardh) Harvey (1849), pl. 236.
Conferva repens J. Agardh (1842), p. 13, non Dillwyn (1802).

Plants form a cushion-like turf, 2–5 cm deep, sometimes covering an area 1 m² in area, or loose lying balls, often blackish green in colour; growth of axes mainly intercalary; fronds densely but irregularly branched, branches often tangled together with rhizoids which may form at base of any axial cell; rhizoids pale green, 50–70 μm diameter; new axes may arise from rhizoids; insertion of branches lateral or subterminal with vertical or near vertical cross walls, eventually two or more branches to one cell; branches sometimes opposite; cells usually cylindrical, 50–200 (–300) μm diameter, 1–12 times as long as broad.

Reproduction mainly by vegetative propagation.

Chromosome no.: not known.

The species occurs in very shaded areas in the lower littoral and in the sublittoral down to 2–3 m, attached to rock or to other algae or lying in loose balls on muddy bottoms of sea lochs; it can also penetrate into estuaries.

A rare plant in the British Isles, but may have been overlooked. It appears to be restricted to the more southerly parts of the country, although van den Hoek (1963) mentions a record from Northumberland based on a specimen in BM (see Batters, 1902); Bardsey Island, Dorset, Devon, Channel Islands, Co. Cork, Co. Wexford.

France, Spain, Portugal, Mediterranean, Adriatic Sea and Black Sea.

Indian Ocean, Atlantic coast of America from Florida south through the Caribbean, Ghana, Cameroon and Solomon Islands to southern Australia.

Plants can be found, apparently, at all times of the year. They have not been recorded in a reproductive condition in the British Isles and spread is probably almost always by vegetative means. Van den Hoek (1963, 1982) has recorded seeing swarmers on two occasions only. The first was in plants from Algiers; here the swarmers appeared to have lost their motility and were developing directly. The second occasion was in plants from Curaçao which were releasing biflagellate swarmers. The life history of *Cladophora coelothrix* is not known.

The plants may grow as a turf of independent plants or may become intertwined forming a very compact turf or cushion; loose lying balls may also be formed in which the branching becomes very irregular and the polarity of the cells inversed. Van den Hoek (1963) gained the impression that cell diameters tended to be greater in localities exposed to wave action than in more sheltered localities.

Feldmann (1937) separated *Cladophora repens* (J. Agardh) Kützing from *C. coelothrix* Kützing on the diameter of the filaments, the former are 150–200 µm and the latter 250–300 µm.

Cladophora dalmatica Kützing (1843), p. 268–269.   Fig. 45

Holotype: L (no. 937. 281. 406; see van den Hoek 1963). Split, Yugoslavia.

Plants 5–10 (–50) cm high, attached by rhizoidal holdfast or forming floating masses, light green to dark green in colour; growth by apical divisions, only later by intercalary divisions at a distance behind apex; branch initials cut off at apical pole of cell by oblique cross walls; branches often falcate or curving backwards, often pectinate, acropetally organised in apical region; branches eventually becoming pseudodichotomous by evection, with up to 6 branches to a cell, sometimes giving the impression of whorled branches; axis sometimes with swellings here and there; apical cells cylindrical with rounded ends, 14–20 µm diameter in light green, unshaded plants, 30–55 (–75) µm diameter in dark green, shaded plants; main axes up to 60–150 µm diameter, often a marked difference in the cell diameters of main axes and branches of higher order.

Reproduction by biflagellate gametes and quadriflagellate zoospores formed in swollen cells of the apical region, formation of gametangia and zoosporangia progressing basipetally; reproduction also by akinetes.

Chromosome no.: n = 12   A. Wik in MS (See Söderström, 1963).
            n = 12 2n = 24 Föyn (1934) for *Cladophora suhriana*, which van
                      den Hoek (1963) regards as *C. dalmatica*. Triploid plants recorded by Föyn (1934) in culture.

Plants occur on both sheltered and exposed coasts, in pools in the upper littoral, in the lower littoral and the shallow sublittoral, epilithic and epiphytic; can form a turf with other low-growing species; also found floating in masses in sheltered bays and lagoons extending into brackish water with salinity down to 24‰ or a little less. According to Söderström (1963) it grows best when water temperature is 18 °C or higher.

Probably more common in the British Isles than indicated by the records because of confusion with other species. Orkney Islands, Sussex, Dorset, Channel Islands, Co. Cork, Co. Galway.

Sweden, Baltic Sea south to the Mediterranean and the Black Sea; the species extends to the tropics on both sides of the Atlantic Ocean: Virginia, Florida, Texas, Bermuda, Bahamas, Jamaica, Puerto Rico, Haiti, Guadaloupe, Curaçao, Columbia; Ghana.

The species is present during the spring and summer months and disappears during the winter. It probably overwinters as basal remnants of plants which later regenerate; it may also survive as undeveloped sporelings. Zoospores and gametes are formed in terminal cells during the summer. The plants are dioecious. The life history, in the cases in which it has been studied, is an isomorphic alternation of gametophyte and sporophyte generations.

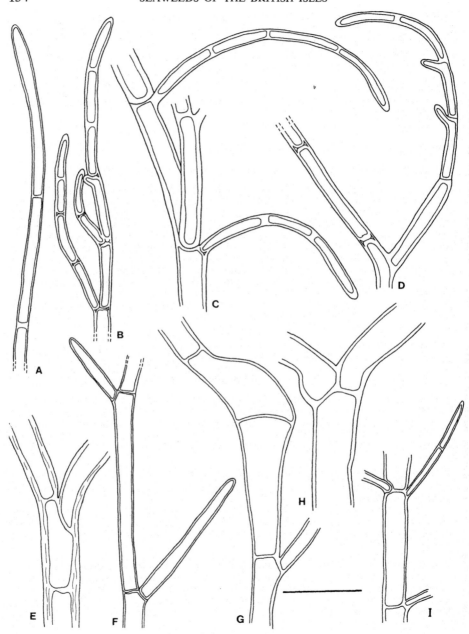

The cells of the thalli of *Cladophora dalmatica* may have irregular swellings and the diameter of the main axes may broaden considerably in contrast to the much narrower lateral branches. Both of these features are described for *C. oblitterata* Söderström, included in *C. dalmatica* (van den Hoek, 1963). Under rough water conditions plants form interwoven carpets less than 5 cm high, or globular tufts in which cells are generally broad and dark green in colour. Under these conditions the branches may be markedly falcate. Under sheltered conditions and high illumination, as in upper shore pools, the plants are longer and much paler in colour, with narrower cells. Plants may become detached and form floating masses comprising delicate branched axes up to 50 cm long.

Cladophora globulina (Kützing) Kützing (1845), p. 219.

Isotype: L (Alg. aq. dulc. germ., Dec. 2, 1833, no. 20; see van den Hoek, 1963). Nr. Tennstedt, Germany.

Conferva globulina Kützing (1833), no. 20.

Plants of long unbranched or sparsely branched filaments attached by rhizoids or unattached; sometimes densely branched in young plants; growth intercalary; branching lateral or subapical, very occasionally pseudodichotomous; apical cell sometimes pointed, cells often slightly swollen at ends; apical cells 13–21.5 μm diameter, 3–8 times as long as broad, filament cells 16–27 (–38) μm, 3–11 times as long as broad.

Vegetative reproduction by fragmentation and by akinetes formed in swollen thick-walled cells, becoming club-shaped.

Chromosome no.: not known.

Mainly a freshwater species in water with a pH value higher than 7, but also occurs in saltmarshes where the salinity may be almost as high as that of sea water, as at the entrance to Poole Harbour, Dorset. Usually found growing in mats of other algae around the bases of larger plants.

Rarely reported in the British Isles; Dorset, Hampshire, Norfolk (Polderman, 1978).

Germany, Netherlands, France, Mediterranean, Black Sea.

Plants are known to be present during December (Polderman, 1978) but no information is available for other periods of the year. Only vegetative reproduction is known for this species and the life history is unknown.

Apart from variations in the degree of branching and the occasional swelling of the cells, little is known about variation in this species. Unbranched filaments can easily be mistaken for a species of *Rhizoclonium* from which they differ, according to van den Hoek (1963) by the higher maximum values for the length/width ratio of the cells and by their occasional apical swellings.

The affinities of this species are perhaps a little uncertain.

Fig. 45   *Cladophora dalmatica*
A–I. Portions of erect filaments. (Lough Ine, Co. Cork; J. Kitching). Bar= 200 μm.

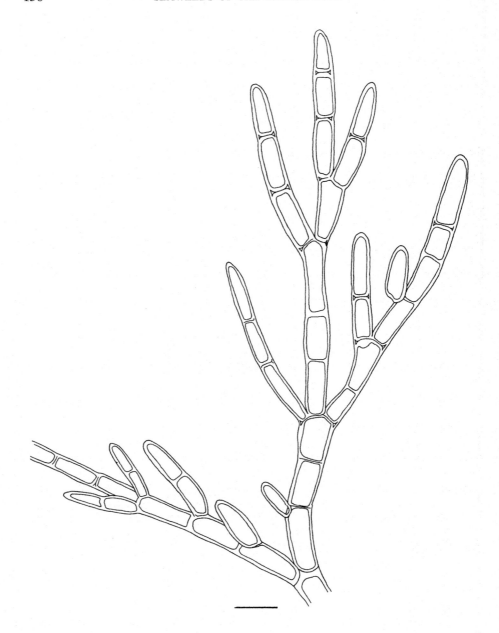

Cladophora hutchinsiae (Dillwyn) Kützing (1845), p. 210. Fig. 46; Pl. 6

Lectotype: BM (No. 8 of H3551/60; designated by van den Hoek, 1963). Bantry Bay, Eire.

Conferva hutchinsiae Dillwyn (1809), p. 65.
Conferva rectangularis Griffiths ex Harvey in Hooker (1833), add. p. X.
Cladophora rectangularis (Griffiths ex Harvey) Harvey (1846), pl. XII.

Plants up to 35 (–40) cm long, slightly coarse in texture, usually attached, but old plants often forming loose-lying masses; growth mainly by intercalary divisions starting just behind the apex; lateral branches usually initially one to each cell, in no definite age sequence, young branches intercalated between older branches, sometimes in opposite pairs, especially in old proliferating axes; branch initials cut off by oblique walls, branching becoming pseudodichotomous by evection; branch base and axial cell becoming decurrent in lower part of plant; eventually two or three branches may form on axial cells; apical cells rounded and blunt (90–) 140–160 (–195) μm diameter, 1–4 times as long as broad, main axes 240–400 μm diameter, with cells 1–3.5 times as long as broad, ultimate branches (100–) 150–170 (–325) μm diameter.

Reproduction by swarmers formed in terminal cells of filaments: release of only biflagellate swarmers so far recorded.

Chromosome no.: 2n = 24 Patel (1961) for plants identified as *Cladophora hutchinsiae* var. *distans* Kützing.

Found in pools from the mid- to lower littoral region and in the sublittoral in both sheltered and exposed localities; absent from brackish waters.

Apart from the specimens checked by van den Hoek (1963) and some checked more recently since the critical characters of the species have become clear, there is uncertainty about the identity of specimens referred to. The distribution given is therefore almost certainly very incomplete.

Ayrshire, Isle of Man, Anglesey, Glamorgan, Dorset, Devon, Cornwall, Isles of Scilly, Channel Islands, Co. Antrim, Co. Londonderry, Co. Mayo, Co. Kerry, Co. Cork.

France, Spain, Portugal, Mediterranean, Adriatic Sea.

The distribution of *Cladophora hutchinsiae* outside southern European waters is uncertain and needs to be checked.

Plants are conspicuous during late spring and the summer months, and can be found as small plants during the winter: plants may also overwinter as denuded bases of branch systems which later proliferate. The reproductive period has not been established. The release of only biflagellate swarmers has been observed and that on only four occasions: twice by Archer (1963) in Anglesey, and by van den Hoek (1963) in the Channel Islands and fertile plants seen by myself in Cornwall during September. The life history is not known.

Fig. 46　*Cladophora hutchinsiae*
Apical region of plant. (Tiree, Hebrides; June 1981; C. A. Maggs). Bar= 100 μm.

Now that the characteristics of this species have been clarified (van den Hoek, 1963), it can easily be recognised by the apical branching pattern, and the broad cells with short length/width ratios. The form of the plants is modified following reproduction which starts at the apex and proceeds basipetally producing barrel-shaped cells which are eventually lost. Axes remaining after swarmer formation continue intercalary divisions and proliferate new branches.

In quiet water, plants floating near the sea bed can have opposite branches as proliferations. According to van den Hoek (1963) this is the form which has been described as *Cladophora rectangularis* (Griffiths ex Harvey) Harvey.

Cladophora laetevirens (Dillwyn) Kützing (1843), p. 267.

Neotype: BM–K (No. H4351/60/6, labelled in Dillwyn's handwriting '1806 L.W.D. *C. laetevirens*'; see van den Hoek, 1963). Nr. Swansea, Wales.

Conferva laetevirens Dillwyn (1809), p. 66, pl. 48.
Conferva falcata Harvey (1849), p. CCVI.

Fronds 10–20 (–30) cm high, light to dark green, forming bushy tufts of well branched fronds; apical division initially giving rise to a distinct acropetal organisation with each axial cell forming a lateral branch; branches cut of by oblique cross walls from apices of axial cells, later forming pseudodichotomies by evection; branches sometimes falcate, often with pectinate arrangement, becoming less regular later following intercalary divisions and new branch formation increasing basipetally; axes never lined with branches of different ages; 1–4 branches per axial cell; apical cells cylindrical with blunt rounded ends, (34–) 45–75 (–100) μm diameter, cells of main axes up to 75–250 μm diameter; cells 2–10 times as long as broad; terminal branch system lost in sporulation followed by proliferation.

Reproduction by quadriflagellate zoospores and biflagellate gametes; zoospores 12–16.5 = 6–7.5 μm; gametes isogamous 7.5–10.5 × 4–6 μm.

Chromosome no.: 2n = about 30 Föyn (1929) for *Cladophora utriculosa*
    2n × 24   Sinha (1958) for a *Cladophora* referred to *C. hutchinsiae*, but which van den Hoek (1963) thought was *C. laetevirens*.

Found in rock pools throughout the littoral region and in the shallow sublittoral in both sheltered and exposed shores; epilithic, epiphytic and epizoic; plants can penetrate estuaries and sea lochs with salinity down to about 25‰.

Widely distributed round the coast of the British Isles from Argyllshire to the south coast; there are so far no definite records north of Jura, Argyllshire.

Argyllshire, Ayrshire, Durham, Isle of Man, Anglesey, Glamorgan, Sussex, Isle of Wight, Dorset, Devon, Cornwall, Scilly Isles, Channel Islands, Co. Mayo, Co. Kerry, Co. Clare.

Fig. 47   *Cladophora lehmanniana*
    A–B. Portion of erect filaments. A. Apical region. (Tiree, Hebrides; June 1981; C. A. Maggs). B. Mid region (Carraroe, maerl beds, Co. Clare; June 1981; C. A. Maggs). Bar= 200 μm.

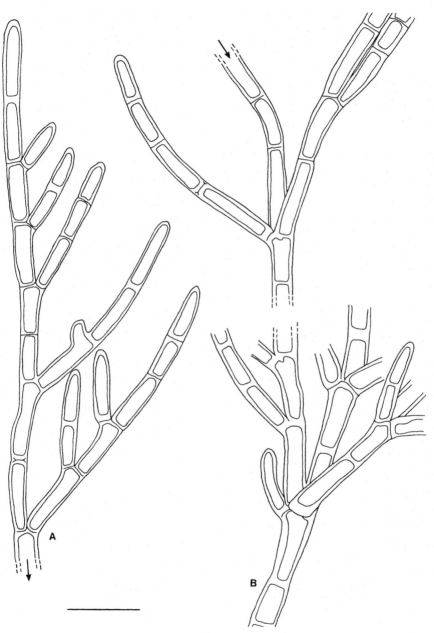

Sweden south to Portugal, Mediterranean, Adriatic Sea, Black Sea.

On the Atlantic shores of North America, mainly in warmer waters of the southern states and the Caribbean (van den Hoek, 1978), eastern Canada (Taylor, 1957; South & Cardinal, 1970; South, 1976); Ghana.

The plants are probably present all the year round, but are most conspicuous during the spring and summer months when they reproduce. Both quadriflagellate zoospores and biflagellate gametes are formed; the plants are dioecious.

The life history is probably an isomorphic alternation of gametophyte and sporophyte generations, but this needs confirmation.

Cladophora laetevirens may form a low-growing spongy turf, 2–4 cm high, of entangled plants mixed with other algal species; branches are then often markedly falcate. At the top of the shore in sunny pools, plants are very slender and light green in colour; lower down the shore and in more shaded situations, the cells are broader and darker green. Plants are reduced to main axes following reproduction which begins at the apices and proceeds basipetally. The remnants of the axes, a form in which the plants may overwinter, later proliferate.

Cladophora lehmanniana (Lindenberg) Kützing (1843), p. 268. Fig. 47

Isotypes: L (No. 937. 281. 47). LD, (Herb. Agardh Nos 9341–9343; see van den Hoek, 1963). Helgoland, Germany.

Conferva lehmanniana Lindenberg (1840), p. 179.

Thallus of sparsely branched axes up to 9 cm long, loose-lying though rhizoids may form at bases, sometimes a single main axis with lateral branches; divisions mainly intercalary; branch initials cut off by oblique, almost lateral cross walls, branching becoming pseudodichotomous by slow evection; older axes may form almost simultaneous proliferations which become separated by intercalary divisions; apical cells 22–40 μm diameter, 5–20 times as long as broad, ultimate branches 22–50 μm diameter, cells 4.5–17 times as long as broad, main axes up to 90($-$100) μm diameter, 2.5–9 times as long as broad.

Reproduction by proliferation and fragmentation; also by akinetes

Chromosome no.: not known.

In pools in the lower littoral and in the sublittoral.

There are few records of this species in the British Isles, though this may be due in part to its confusion with other broad-celled *Cladophora* species. It belongs to the warmer waters of the Mediterranean, but reaches the south coast of England and the west coast of Ireland.

Devon, Isle of Wight, Co. Galway.

Germany, France, Spain, Portugal, Mediterranean and Adriatic Sea.

It is not recorded outside this southern part of the European Atlantic coast.

The plants are known to be present during the summer, but the seasonal pattern of growth is not known, nor is there any information on its reproductive period. Van den Hoek (1963) found plants from Spain producing quadriflagellate zoospores on one occasion.

The life history is not known.

There is little information on form variation for this species. As is the case for *Cladophora hutchinsiae*, old axes which have lost their apices form proliferations which often have opposite branches; such plants have been referred to *Cladophora rectangularis* (Griffiths ex Harvey) Harvey (see *C. hutchinsiae*, pp. 157, 158).

Cladophora liniformis Kützing (1849), p. 405.   Fig. 48

Holotype: L (No. 937. 281. 32; see van den Hoek, 1963). Venice, Italy.

Plants up to 35 cm high, somewhat stiff, dark green, of richly branched fronds; growth of axes mainly by apical segmentation, giving an acropetal sequence of branches, intercalary divisions beginning well behind apex; branch initials cut off by oblique walls, evection later producing pseudodichotomies, eventually 1–3 branches on an axial cell, branches sometimes opposite; apical branch system often falcate or

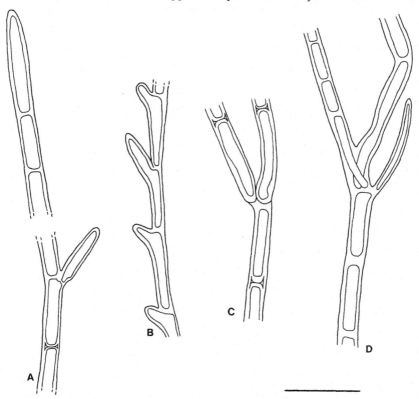

Fig. 48   *Cladophora liniformis*
A–D. Portions of erect filaments. (The Wash). Bar= 200 μm.

refractofalcate; apical cells cylindrical with rounded ends, (90–) 114–120 (–160) μm diameter, 2–6 (–10) times as long as broad, main axes 140–330 μm diameter, 2–10 times as long as broad.

Reproduction by quadriflagellate zoospores, (13–) 15–16.5 (–18) × 8 (–10) μm, rarely formed.

Chromosome no.: not known.

Found lying on mud and sand or free-floating in estuaries, harbours and lagoons and in ditches by the sea, often entangled with *Enteromorpha* and other algae around the bases of *Zostera* plants: the plants belong to essentially marine habitats.

The species may be more common in the British Isles than the few records suggest. Norfolk, Glamorgan, Hampshire, Channel Islands.

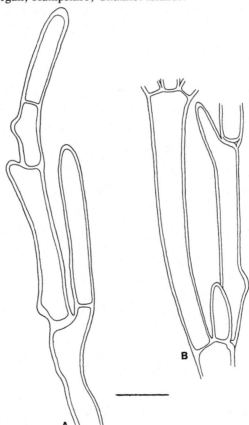

Fig. 49   *Cladophora pellucida*
A, B. Portions of erect filaments. (Mevagissey, Cornwall; April 1981). Bar= 200 μm.

Arctic Russia, Germany, Netherlands, France, Mediterranean, Adriatic Sea, Black Sea.

The Atlantic coast of North America from Quebec south to the Caribbean.

The plants are probably present all the year round but information on their occurrence is scanty. They probably always reproduce vegetatively by fragmentation and proliferation; akinetes are sometimes formed.

The plants vary greatly in the degree and form of branching due to irregular proliferation.

Cladophora pellucida (Hudson) Kützing (1843), p. 271.   Fig. 49

Neotype: BM (Designated by van den Hoek (1963) as one of three specimens 'preserved under the name *Conferva pellucida* in Forster's herbarium labelled "Hudson's sale"'). Walney Island, see Archer (1963).

Conferva pellucida Hudson (1762), p. 483.
Cladophora pseudopellucida van den Hoek (1963), p. 218.

Thallus up to 30 cm, but usually less than 10 cm high, pure translucent green to dark green, rather wiry texture, with marked dendroid branching system arising from single or few axes, attached by rhizoids arising from basal cell; rhizoids form a complex system from which erect fronds may arise; growth by strictly apical division giving an acropetal organisation of branches with single-celled internodes; branch initials cut off at an acute angle by oblique cross walls, becoming pseudodichotomous by evection, eventually 2–4 (–6) branches on each axial cell, appearing verticillate lower down axis; apical cells (55–) 90–255 µm diameter, 1–12 (–20) times as long as broad; main axes 250–500 µm diameter, lower cells somewhat club-shaped, cells becoming longer basipetally, basal cell (250–) 400–600 µm diameter, up to 20 or more times long as broad.

Reproduction by quadriflagellate zoospores and biflagellate gametes formed in terminal cells of filaments, preceded by intercalary division and proceeding basipetally; plants dioecious.

Chromosome no.: uncertain; n = 18, 2n = 33 ? Sinha (1958).
n = about 16, 2n = about 32 Föyn (1929).

Occurs in the lower littoral on rocky shores in deep pools, seldom exposed by the tide; also in the sublittoral; it occurs on semi-exposed shores, often under overhangs, but frequently also in pools exposed to bright light as well as in those that are shaded.

The species belongs to the more southern parts of the British Isles.
Isle of Man, Lincolnshire, Bardsey Island, Anglesey, Glamorgan, Lundy Island, Kent, Isle of Wight, Dorset, Devon, Cornwall, Channel Islands, Co. Mayo, Co. Galway, Co. Down.
France, Spain, Portugal, Mediterranean, Adriatic Sea, Canary Islands.
Apparently limited to the southern European Atlantic coast and the Mediterranean.

Present throughout the year, but most conspicuous during the winter months. Vegetative growth occurs through the whole year. Little is known about the reproductive seasons of this species: Archer (1963) observed the release of zoospores on one occasion, though the date is not recorded.

The life history is probably an isomorphic alternation of gametophyte and sporophyte generations.

There appear to be no marked changes in morphology of the plants related to seasons. The plants vary in size and degree of branching and in the size of their cells. A small-celled form was separated by van den Hoek (1963) as *Cladophora pseudopellucida*, but later he decided, on the basis of culture experiments that it and *C. pellucida* are conspecific (van den Hoek, 1978).

The species is readily recognised by its long translucent segments, particularly marked in the mid- and lower sections of the plants. Basal filaments are often coloured red due to the presence of *Schmitziella endophloea* Born. & Batt.

Cladophora prolifera (Roth) Kützing (1843), p. 271.   Fig. 50

Neotype: L (937. 264. 23 designated by van den Hoek, 1963, p. 208). A specimen collected by Hauck at Miramare, Trieste.

Conferva prolifera Roth (1797), p. 182–183.

Thallus consisting of dark green interwoven tufts of fronds becoming dark brown on drying, 5–10 (–25) cm high, rather rigid; growth by apical divisions giving a very regular acropetal organisation of branches, each cell forming one, later two or three branches, arrangement sometimes pectinate; branch initials cut off by almost vertical cross walls, evection gradually forming pseudodichotomies; lower cells of axes form branched rhizoids with annular constrictions, rhizoids winding around lower axial cells and growing down over substratum; apical cells cylindrical with rounded apices, straight or slightly curved, 90–200 µm diameter; main axial cells elongated, club-shaped, 330–650 µm diameter, up to 10–18 times as long as broad.

Reproduction by biflagellate swarmers formed in cells of apical region; sporangial formation preceded by intercalary division.

Chromosome no.: not known.

Grows in the lower littoral region on steep shaded rocks and in rock pools, and in the sublittoral down to a depth of 20–30 (–70) m. It belongs to sheltered or very slightly exposed shores: although it is normally found in shaded places, it can grow in full sunlight in deeper waters.

The species belongs mainly in the Mediterranean, but has been found in the British Isles; it is a rare alga here, being recorded from only three localities.

Dorset: a specimen was found in the Fleet, Weymouth, in 1976. Previous records from Weymouth are indicated by specimens provided by Holmes in 1884 (Alg. Brit. Rar. No. 32); Cotton (1912) says that it was washed ashore in Weymouth in profusion in 1884 and 1885.

Fig. 50   *Cladophora prolifera*
A–D. Portions of erect filaments. Note production of rhizoidal filaments from basal region cells in D. (The Fleet, Dorset; Aug. 1976; W. F. Farnham). Bar= 500 µm.

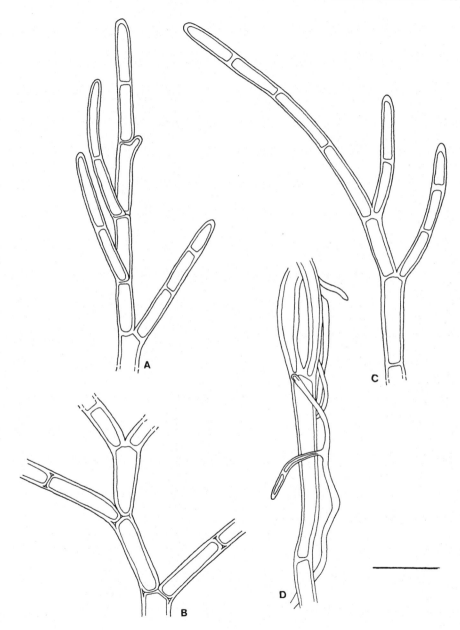

Colonsay, Inner Hebrides (Norton *et al.*, 1969), Clare Island, Co. Mayo (Cotton, 1912). Dickinson (1963) decided that the Clare Island specimens were stunted *C. pellucida*.

Atlantic coasts of France and Spain, Mediterranean, Adriatic Sea.

Azores, Atlantic coast of North America from northern Carolina to the Caribbean.

Plants are probably present throughout the year, but most of the records refer to the summer months. Hamel (1924) reported the species to be more abundant in summer than in winter in the Gulf of Gascoigne while Funk (1955) found young plants in August and September at Naples. Production of swarmers has not been reported for the British Isles. Van den Hoek (1963) found biflagellate swarmers on a few occasions in plants from France and Spain and he reported that they could germinate directly to form new plants.

The life history is not known.

According to van den Hoek (1963) the morphology of the plants of this species is very constant, but too few specimens have been seen from the British Isles to comment on this.

Cladophora pygmaea Reinke (1888), p. 241.   Fig. 51

Lectotype: KIEL (in herb. Reinke; see van den Hoek, 1963). Strand, Kieler Förde, Germany.

Plants 1.0–1.5 mm high, well branched and somewhat stiff, attached by a disk holdfast formed by the expanded base of the single elongated basal cell, occasionally lobed; growth of axes mainly intercalary; branching irregular, sometimes subsecund; branch initials cut off by oblique cross walls, axes becoming pseudodichotomous by evection; cells barrel-shaped (25–) 39–65 (–75) μm diameter, 29–111 μm long; cell walls thick and striate.

Reproduction by quadriflagellate zoospores formed in chains of barrel-shaped cells in upper part of frond; zoospores released through pore midway along cell.

Chromosome no.: not known.

Occurs on small stones, pieces of shell and pottery and in crevices in the sublittoral along with encrusting red and brown algae and in maerl beds, down to a depth of 20–30 m or more; it is reportedly encouraged by strong water currents; plants may grow in brackish water in the Baltic Sea.

The British records of *Cladophora pygmaea* date from 1975 and result from the diving activities of Irvine *et al.* (1975), Maggs & Guiry (1981), Maggs, Freamhainn & Guiry (1983).

Shetland, Hebrides, Isles of Scilly, Co. Down, Co. Donegal, Co. Galway, Co. Clare.

Norway, Baltic Sea, Germany, France (Brittany).

The Atlantic coast of North America from Newfoundland south to Massachusetts.

Plants appear to be present throughout the year. The reproductive period is not known for the British Isles, but plants forming swarmers have been found in September in Brittany (Maggs & Guiry, 1981), in July and August in Newfoundland

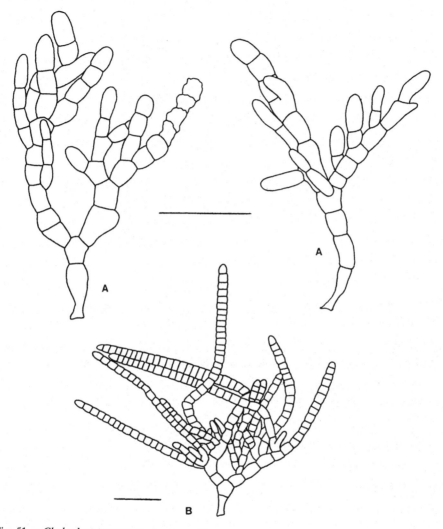

Fig. 51   *Cladophora pygmaea*
A-B. Habit of plants attached at the base by a single, elongated cell. Figure drawn by
G. Russell. Bar= 250 μm (A), 200 μm (B).

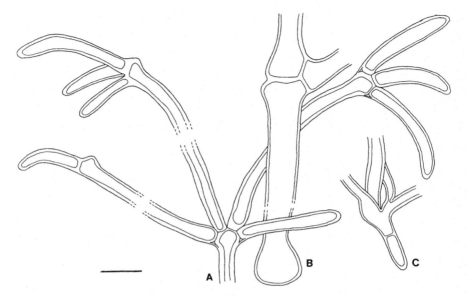

Fig. 52   *Cladophora retroflexa*
A–C. Portions of erect filaments (Carraroe, maerl beds, Co. Clare; May 1981). Bar= 500 μm.

(Hooper & South, 1977), in July and possibly a little earlier in Massachusetts (Wilce, 1970). The behaviour of the zoospores and the life history are not known.

Cladophora retroflexa (Bonnemaison ex Crouan frat.) van den Hoek (1963), p. 215.   Fig. 52

Isotype: L (Desmazieres, Pl. crypt. France, ed. 2, sér. 1, No. 868, without description; see van den Hoek, 1963). Brittany, France.

Conferva retroflexa Bonnemaison ex Crouan frat. (1867), p. 127.

Plants dark green, unattached, forming aegagropiloid tufts or balls; rhizoid-like branches sometimes formed if cells are lost at the base; growth by divisions of apical cells; branches formed acropetally and often incurved, arising laterally with almost vertical cross walls; several branches arising later on cells in lower part of plant; cells sometimes slightly club-shaped at base; apical cells 100–175 (–265) μm diameter, 2–6 times as long as broad; basal cells up to 170–220 μm diameter, 6–12 times as long as broad.

Reproduction by vegetative propagation only.

Chromosome no.: not known.

Forming aegagropiloid masses tangled around the bases of other algae and around

Zostera plants in sheltered bays and sea lagoons; found in the lower littoral and shallow sublittoral.

There are only two recent records of this species in the British Isles, from Portland Harbour in Dorset and from Co. Galway; the previous records were from the same two sites, from Weymouth in 1884 and Co. Galway in 1899 (see Batters, 1902). The Atlantic coast of France and the Mediterranean and Adriatic Seas.

The British records are for the summer months, but the plants may be present all the year round. Swarmers have not been seen for this species and the life history is not known.

Cladophora rupestris (Linnaeus) Kützing (1843), p. 270.   Fig. 53; Pl. 7

Lectotype: OXF (in Dillenius Herb.; see van den Hoek, 1963). Probably Bognor, Sussex.

Conferva rupestris Linnaeus (1753), p. 167.

Plants up to 15–20 cm high, dark often bluish green in colour, somewhat stiff texture, often rope-like in form; basal plate of rhizoids giving rise to numbers of erect fronds; fronds straight or slightly curved outwards; growth mainly by intercalary divisions, forming axes lined with branches of different ages, younger branches intercalated between older ones; branch initials cut off by oblique cross walls at an acute angle, axes becoming pseudodichotomous by evection, eventually 3–4 (–6) branches on an axial cell; cells decurrent at insertion of branches in lower part of plant; apical cells slightly tapering, 40–80 μm diameter, 2–6 times as long as broad; cells of main axes up to 90–220 μm diameter, 1.5–7 times as long as broad.

Reproduction by quadriflagellate zoospores and biflagellate isogametes formed in terminal cells of fronds, usually following intercalary divisions; plants dioecious; zoospores 17–25 × 11–15.5 μm, gametes 10.5–15 × 7.5–10.5 μm.

Reproduction in the juvenile state of developing zoids has been shown by Archer (1963) and van den Hoek (1963).

Chromosome no.: n = 12, 2n = 24 (Archer, 1963; Sinha, 1958; Patel, 1961)
n = 6, 2n = 12 (Jönsson & Perrot, 1967 for material from France).

The plants grow in pools, on the rock surface, hanging in ropes in crevices or forming undergrowth to larger algae at all levels of the shore. They sometimes form an almost complete cover of short, usually stunted plants in rock pools at high tide level and occasionally can be found in the splash zone of the supralittoral; they occur on both sheltered and exposed coasts, are more luxuriant and possibly more typically developed under more exposed conditions. Plants may also be epiphytic.

Common on all parts of the British coast.

European Atlantic coast from Scandinavia to the Mediterranean, Adriatic, Baltic Sea, Murman Sea, White Sea. The most northerly site is probably on the west coast of Novaya Zemlya (Kjellman, 1883).

Atlantic coast of North America from the Canadian Arctic south to Massachusetts, Greenland; Iceland, Faroes.

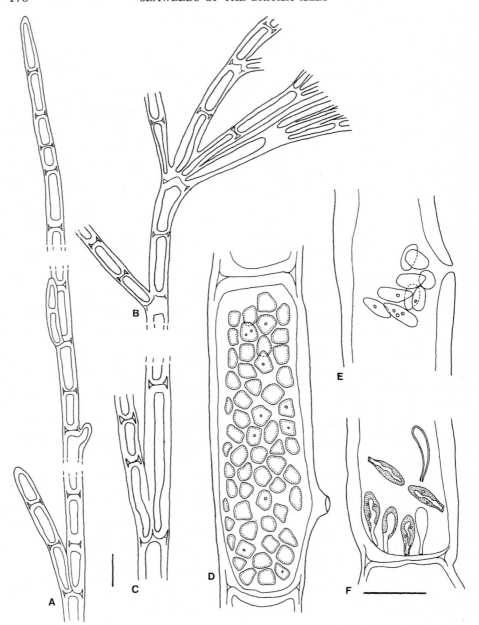

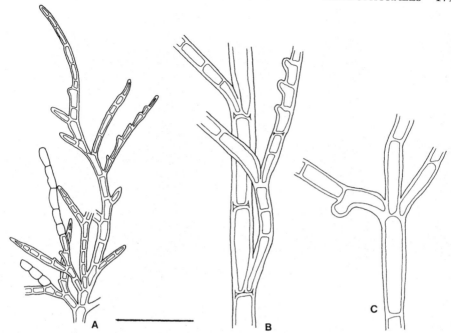

Fig. 54   *Cladophora sericea*
     A–C. Portions of erect filaments. (Berwick). Bar= 200 μm.

The species occurs throughout the year at all levels of the shore and in the sublittoral, attaining maximum development in summer near low tide level. Both zoospores and gametes can be found at most times of the year. Archer (1963) was unable to find any correlation between times of reproduction, the state of the tide or environmental conditions.

The life history is probably an isomorphic alternation of gametophyte and sporophyte generations; the plants are dioecious.

The morphology of the plants is fairly constant over a wide range of habitat conditions and over a wide geographical area. Form range is affected by physical damage due to grazing by animals and by loss of the apical region on reproduction, both processes followed by regeneration and proliferation of branches. Plants sometimes form an almost complete cover of stunted growth at high tide level and occasionally in the splash zone where pools are brackish. Filaments are short and branching dense in the most exposed situations. The species has a wide salinity tolerance, sometimes down to about 5‰.

Fig. 53   *Cladophora rupestris*
     A. Slightly curved apical segment. B. Mid region segment. C. Basal region segment.
     D–F. Reproductive cells. Bar= 100 μm (A–C), 50 μm (D–F).

Cladophora sericea (Hudson) Kützing (1843), p. 264.   Fig. 54

Holotype: OXF (Marine material of *Conferva trichodes virgata, sericea* in Dillenius's herb.; see van den Hoek, 1963). Sheerness, Kent.

Conferva sericea Hudson (1762), p. 485.
Cladophora rudolphiana (C. Agardh) Kützing (1843), p. 268.
Cladophora glaucescens (Griffiths ex Harvey) Harvey (1849), pl. 196.
Cladophora flexuosa (O. F. Müller) Kützing (1843), p. 270.

Plants 5–25 cm high, pale to dark green, richly branched; growth of axes mainly by intercalary divisions, new cells giving rise to rows of branches of mixed ages, younger branches intercalated between older ones; branch initials cut off by oblique cross walls, axes becoming pseudodichotomous by evection; strong tendency for cells at bases of branches to become decurrent with axial cells; apical cells and young branches distinctly tapering, 18–24 μm diameter in pale green plants, 50–70 μm diameter in dark green plants; main axes 55–170 μm diameter.

Reproduction by quadriflagellate zoospores and biflagellate isogametes formed in apical region of plants following intercalary divisions; zoosporangia sometimes in long terminal chains of barrel-shaped cells; zoospores (13.5–) 17–24 (–27) × (6–) 8–12 (–14) μm, gametes (8–) 10–16 (–18) × (4–) 5–8 (–9.5) μm.

Chromosome no.: n = 12, 2n = 24 Schussnig, 1939, 1960 (for *Cladophora gracilis*),
        Sinha 1958 (for *Cladophora flexuosa*).

2n = 24 Patel (1961), but Söderström (1963) thought that the plant involved, in BM, belongs to *Cladophora laetevirens*.

2n = 22 Wik in Söderström (1963) (for *Cladophora glaucescens*).

In rock pools or on the rock surface over the whole littoral region and in estuaries and lagoons with salinities down to about 15‰, on both exposed and sheltered shores; often an undergrowth species.

Found all round the coast of the British Isles from the Shetland Isles to the Scilly Isles and the Channel Islands.

From the European Arctic to the Mediterranean, Adriatic Sea, Black Sea, Baltic Sea, White Sea.

Atlantic coast of North America from the Canadian Arctic south to Georgia; Iceland, Faroes; Pacific coast of North America from Alaska to North Washington; Japan; Australia.

The plants grow during the spring and summer months and reproduce during the summer. Following reproduction, the apical portions of the plants are lost and the remnants of the main axes survive the winter, proliferating new axes during spring. Both quadriflagellate zoospores and biflagellate gametes are formed on separate plants and the gametophytes are dioecious.

––––––––––––

Fig. 55   *Cladophora vagabunda*
       A–C. Portions of erect filaments. (H. Blackler; Dec. 1978). Bar= 200 μm.

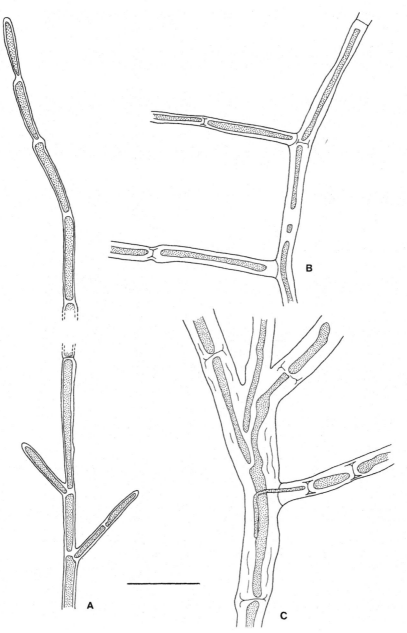

The life history is an alternation of gametophyte and sporophyte generations (Bliding, 1936) or there may be only an asexual cycle involving biflagellate swarmers which germinate directly to form new plants (Bliding, 1936; van den Hoek, 1963). Van den Hoek (1963) has separated plants with an alternation of generations, as *C. sericea* (Huds.) Kütz. var. *sericea* Hoek, from those with an asexual cycle, as *C. sericea* (Huds.) Kütz. var. *biflagellata* Hoek.

Under exposed conditions and in turfs of mixed species as undergrowth to larger intertidal algae, the plants may be less than 5 cm high and the apical branches somewhat refracted. Under sheltered conditions in rock pools, the plants are much longer and the apical branches straight. Under conditions of high illumination, cells of the axes and branches are narrower and lighter in colour than the broad darker green cells formed under shady conditions.

Cladophora vagabunda (Linnaeus) van den Hoek (1963), p. 144.   Fig. 55

Lectotype: OXF (*Conferva marina trichodes lanae instar expansa* in Herb. Dillenius, designated by van den Hoek, 1963). Selsey, Sussex.

Conferva vagabunda Linnaeus (1753), p. 1167.

Plants up to 50 cm long, consisting of well branched axes, attached by rhizoidal holdfast, or free-floating; apex of axis with regular acropetally organised branch system, often falcate or with degree of refraction; growth mainly apical, each new cell giving rise to a lateral branch, intercalary divisions beginning distant from the apex producing secondary branches; branches obliquely inserted at cross walls, becoming pseudodichotomous by evection; in floating plants branching often wide-angled, up to 5 branches eventually formed on axial cells; vegetative apices often slightly tapering, becoming rounded or club-shaped during reproduction; apical cells 17–85 μm diameter, 1.5–23 times as long as broad, main axes up to 80–300 μm diameter, 1.5–14 times as long as broad.

Reproduction by biflagellate isogametes and by quadriflagellate zoospores; gametes (7.5–) 10–12 (–15) × (4.5–) 6–8 (–11) μm, zoospores (11–) 16–20 (–22.5) × (6–) 8–9 (–14) μm: reproduction also by akinetes.

Chromosome no.: not known.

Found growing in rocky tide pools from the upper to the middle littoral region on fairly sheltered shores; may be attached to small stones, shells and other algae, later forming dense floating masses in quiet shallow sea lagoons and in estuaries with muddy bottoms; plants may lie on muddy bottoms of pools and ditches in salt marshes; the species is essentially marine, but can penetrate into waters with salinities down to at least 15‰.

Infrequently recorded in the British Isles. It has probably been confused with *Cladophora glomerata* and may be more common than the records indicate.
     Dorset, Sussex, Norfolk, Outer Hebrides, Co. Antrim, Co. Clare.
     Southern Scandinavia and the Baltic Sea south to the Mediterranean, Black Sea and Caspian Sea and to the tropics; Liberia, Ghana, Benin, W. Cameroon.
     On the Atlantic coast of North America from Nova Scotia through the Caribbean to Rio de Janeiro.

The species is probably present all the year round with its main growth during spring and summer. In winter the plants are reduced to basal axes which become thick-walled and starch-filled, following reproduction; such denuded axes later proliferate new branch systems. Swarmers have been recorded in late summer.

The life history is not completely known, but the finding of plants producing isogametes or quadriflagellate zoospores suggests that there is an isomorphic alternation of generations. The plants are dioecious.

Whether growing on rock surface or on a more mobile substrate, such as small stones and shells, plants frequently become detached and free-floating at the water surface or on muddy substrates. Under bright illumination apical axes and branches are narrow and pale in colour, while more shaded plants or parts of plants are broader, within the limits for the species, and much darker green. Plants become reduced to main axes following loss of tissue on release of swarmers. Such axes usually proliferate acropetally organised branch systems; the new branches may be opposite giving the plants the appearance of *Cladophora rectangularis* (Griff. ex Harv.) Harv. (see *C. hutchinsiae*, pp. 157, 158).

RHIZOCLONIUM Kützing.

RHIZOCLONIUM Kützing (1843), p. 261.

Type species: *Rhizoclonium juergensii* (Mertens) Kützing (1843), p. 261 (= *R. tortuosum* (Dillwyn) Kützing, 1845, p. 205). See Koster (1955).

Thallus consisting of a tangle of prostrate unattached uniseriate cellular filaments, unbranched, though often with simple or branched unicellular, occasionally multicellular rhizoids; branches similar to parent filaments, very rare; cells multinucleate with reticulate chloroplast and several pyrenoids; mean cell width 6–27 μm.

Reproduction in the field mainly vegetative by fragmentation and by formation of akinetes; also by quadriflagellate zoospores and biflagellate gametes.

The genus is world-wide in its distribution and many species have been described. Of the three species, *Rhizoclonium arenosum* (Carm. ex Harv. in Hook.) Kütz., *R. lubricum* Setch. & Gardn. and *R. riparium* (Roth) Harv. listed by Parke & Dixon (1976), only *R. riparium* (= *R. tortuosum* (Dillw.) Kütz.) is accepted here. The only British record of *R. lubricum* is that of Christensen (1957) from the Goleen, Lough Ine, Co. Cork. He gave a mean cell width of 48.3 μm, a value that falls within the variation pattern for *Chaetomorpha capillaris* in Price's (1967) scheme. Price, who looked for the species in the Goleen in 1964, was unable to find it. It has not been found possible to distinguish *R. arenosum* and this species is shelved pending further investigation.

Only one species in the British Isles.

Rhizoclonium tortuosum (Dillwyn) Kützing (1845), p. 205. Fig. 56

Conferva tortuosa Dillwyn (1805), pl. 46.
Conferva riparia Roth (1806), p. 216.
Rhizoclonium riparium (Roth) Harvey (1849), pl. 238.
Conferva implexa Dillwyn (1809), p. 46.

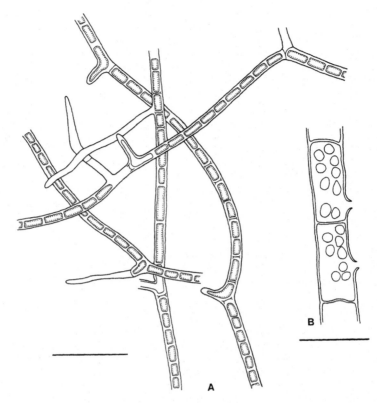

Fig. 56　*Rhizoclonium tortuosum*
A. Entangled filaments with short, lateral rhizoids. (Orkney). Bar= 200 μm. B. Reproductive cells. (Aberdeen). Bar= 50 μm.

Rhizoclonium implexum (Dillwyn) Kützing (1845), p. 206.
Rhizoclonium juergensii (Mertens) Kützing (1843), p. 261.
Rhizoclonium lacustre Kützing (1849), p. 385.
Rhizoclonium affine Kützing (1849), p. 385.
Rhizoclonium biforme Kützing (1849), p. 384.
Rhizoclonium interruptum Kützing (1845), p. 205.
Rhizoclonium kochianum Kützing (1845), p. 206.
Rhizoclonium kerneri Stockmayer (1898), p. 583.

Thallus of unattached filaments forming entangled fleece over substrate; texture soft and flaccid; plants pale to dark green; rhizoids present or absent, if present few to numerous, septate or non-septate, septate rhizoids up to 5 cells long; filaments of cylindrical cells, mean width range 6–27 μm, cells 1–8.5 times as long as broad; cell walls 2–3 μm thick, not striated, not constricted at septum; chloroplast parietal, reticulate with several pyrenoids, 2–9 nuclei arranged axially.

Reproduction by quadriflagellate zoospores, 12.5–18.1 µm long and biflagellate isogametes, 9.1–13.2 µm long; vegetative reproduction by fragmentation and by akinete formation.

Chromosome no.: $n = 18$ $2n = 36$ Patel (1961) for *Rhizoclonium riparium*.
$2n = 24$ Patel (1961) for *R. implexum*.

Found growing on mud, sand and rock and in crevices in cliff faces, in the upper littoral region, sometimes associated with salt marsh conditions.

Distributed round the whole coast of the British Isles.

Found from Scandinavia south to France and the Mediterranean, Baltic Sea, Adriatic Sea, Canary Islands.

The species is probably more or less world-wide in its distribution. West coast of America from Alaska to Chile; east coast from Labrador to the Caribbean; West Africa; Persian Gulf, Indian Ocean, Japan, Australia, New Zealand.

Plants are abundant in spring, summer and autumn, less so in winter. Swarmers appear to be formed only rarely in the field. Price (1967), in three years of sampling at monthly intervals in Anglesey, Wales, only once found swarmers, a few dead ones in an almost empty cell. Material collected by Bliding (Bliding, 1957) in Anglesey in July 1953, produced quadriflagellate swarmers after 10 weeks in culture. He found the plants to have an isomorphic alternation of generations.

It appears that the alternation of generations is a rare occurrence in the field in the British Isles and it may be that it is to a large extent replaced by vegetative reproduction.

Price (1967) found that individual samples of *Rhizoclonium tortuosum* showed a range of variability in mean cell width that was greatest during winter, spring and the early summer months and least between July and November. Cell length values also showed a seasonal variation with minimum mean values in July increasing to maximum values in January and December, suggesting that cell division is more active in late summer than at other times of the year. Mean cell width and mean cell length values vary also on a geographical basis within the limits given for the species, but there does not seem to be any distinct pattern. The differences could be due to local conditions or might be genetically based.

The conspecificity of *Rhizoclonium tortuosum* (= *R. riparium*) and *R. implexum* has already been referred to in Parke & Dixon (1976): see Nienhuis (1975). The systematic position of the plant described by Perrot (1965) under the name *Lola implexa* (Dillwyn) G. Hamel (1931*b*; Harvey, 1846, pl. LIV b), with a life history involving only diploid plants reproducing by anisogamous gametes and biflagellate zoospores and with a chromosome number of $2n = 20$, is uncertain. This has not been recorded for the British Isles.

WITTROCKIELLA Wille

WITTROCKIELLA Wille (1909), p. 16.

Type species: *Wittrockiella paradoxa* Wille (1909), p. 16.

Thallus of minute branched multicellular filaments, loosely associated or densely interwoven, forming mats or balls; filaments sparsely or more richly branched, erect, surrounded by mucilage, forming unicellular hairs above, rhizoidal branches below, sometimes with a well branched prostrate system; cells multinucleate, dark or light to yellow-green; chloroplast parietal, reticulate, with numerous pyrenoids.

Reproduction by akinetes and aplanospores formed in upper filament cells.

Wittrockiella, originally placed by Wille (1909) in a separate family Wittrockiellaceae, is a slightly anomalous member of the Cladophoraceae. In many ways it is very similar to the *Aegagropila* section of the genus *Cladophora* (see van den Hoek, 1963, 1982), but resembles some members of the Chaetophoraceae in the occasional production of long unicellular hairs with swollen bases. Since first being found by Wille (1909) in Norway, the genus has only recently been rediscovered in Europe (Poldermann, 1976). It is now known to occur in both northern and southern hemispheres (see Chapman, 1949; van den Hoek, Ducker & Womersley, 1984).

Only one species in the British Isles.

Wittrockiella paradoxa Wille (1909), p. 16.

Type: O? Lyngør, Norway.

Thallus of branched filaments, sometimes *Cladophora*-like, often golden in colour, with rhizoidal branches; vegetative branches arising from upper ends of cells, sometimes with very long hairs; cells spherical, ovoid or elongate 20–40 (–50) μm diameter, 80–250 μm long; chloroplast filling cell, reticulate with several pyrenoids; rhizoidal branches up to 1 mm long, cells 8–20 μm diameter, 10–100 μm long, chloroplast small, compact, parietal with 1–4 pyrenoids, sometimes reticulate.

Reproduction by akinetes formed from upper filament cells, 60 (–100) μm diameter, formed singly or in series, cell wall thick and layered; or by spherical aplanospores, 5–10 μm diameter, formed in swollen upper cells of filament.

Chromosome no.: not known.

Found growing over the surface of salt marshes in mats of other algae and epiphytic or partially endophytic on stems of salt marsh plants such as *Halimione portulacoides*, also on species of *Zostera*.

Rare in Britain, but possibly overlooked. Norfolk, Hampshire, Dorset, Argyllshire. Norway, Netherlands.
Massachusetts, Washington, USA.

Probably present all the year round but records are incomplete. According to Poldermann (1976) it is most frequently found in spring on salt marsh substrates, but it has no particular seasonal periodicity as an epiphyte.

The life history is unknown and there is too little information to make an assessment of its seasonal reproductive behaviour.

In plants growing on 'soil', the filaments are erect and generally unbranched or with few short branches. As epiphytes or partial endophytes on salt marsh plants, the plants consist mainly of rhizoidal branches with few erect vegetative cells (Poldermann, 1976).

CODIALES Feldmann

CODIALES Feldmann (1954), p. 97.

Siphonales Wille in Warming (1884), p. 33.
Derbesiales Feldmann (1954), p. 97.

Thallus coenocytic, a globose vesicle or series of connected vesicles, or branched filaments, filaments erect and/or prostrate, free or organised into a solid spongy plant body of definite shape, shape various; cross walls in filaments occasional, mainly in formation of zoosporangia and gametangia, vegetative branches occasionally cut off by double cell wall enclosing small cell; thallus uncalcified or calcified; chloroplasts small, discoid, oval or spindle-shaped, with or without pyrenoids.

Sexual reproduction by isogamous or anisogamous biflagellate gametes; plants dioecious or monoecious; asexual reproduction by stephanospores; vegetative reproduction by fragmentation, regeneration of isolated branches and isolated cytoplast, or by aplanospores.

The life history, where known, is a succession of diploid generations or a heteromorphic alternation of gametophyte and sporophyte generations.

The Order Codiales was created by Feldmann (1954). He included in it both Codiaceae and Bryopsidaceae on the grounds that the plants involved were homoplastic and were diplonts. He excluded *Derbesia* which he placed in a separate Order, Derbesiales, because, though homoplastic, it had been shown to have a heteromorphic type of life history. It is now known that, on the basis of structure and life history, *Bryopsis* and *Derbesia* are so closely similar and grade into one another that they must be placed in the same order and family and some authors (e.g. Neumann, 1964) would suggest even in the same genus. Recent work (Kermarrec, 1980) has shown that *Bryopsis* is a diplont despite the fact that it has two heteromorphic reproductive phases and thus differs from *Derbesia*. On this basis there seems no justification for merging the two genera. The third family, Chaetosiphonaceae, included in the Codiales, contains two genera, the systematic position of both is still uncertain.

The families within this order are: Bryopsidaceae, Codiaceae, Chaetosiphonaceae.

BRYOPSIDACEAE Bory orth. mut. De Toni

BRYOPSIDACEAE De Toni (1888), p. 449.

Bryopsideae Bory (1829), p. 203.
Derbesieae Thuret in Le Jolis (1863), p. 14.
Halicystaceae G. M. Smith, (1930), p. 227.

Thallus of uncalcified, branched, coenocytic filaments of either unlimited or limited growth, or a multinucleate unicellular vesicle; filaments not woven into solid thallus of definite shape; filaments with central vacuole surrounded by cytoplasm with disk- or spindle-shaped chloroplasts, with or without pyrenoids, many small nuclei; cross walls formed only in separation of reproductive structures, sometimes also at bases of vegetative branches; reproduction by anisogamous biflagellate gametes and by stephanokont zoospores; vegetative reproduction by fragmentation, development of isolated pieces of cytoplast and by aplanospore formations.

The Bryopsidaceae forms a fairly coherent family now that the relationships between species of the genera *Bryopsis* and *Derbesia* have been clarified by experimental work. The finding of stephanospores in the life history of species of *Bryopsis* (Rietema, 1969, 1970, 1971a, b) and the linking of plants with the morphology of *Derbesia* in the life histories of *Bryopsis* (see Neumann, 1974; Rietema, 1975) have indicated a close relationship between the two genera. It has been known for some time that species of the genus *Halicystis* belong in the life histories of species of *Derbesia* (Feldmann, 1950, 1952; Kornmann, 1938; Neumann, 1969b; Page, 1970). *Ostreobium* Bornet & Flahault has been included in the family though its systematic position is still uncertain. A phase with the general morphology of *Ostreobium* has been found in the life history of *Pseudobryopsis myura* (Mayhoub, 1974a, b) though Kornmann & Sahling (1980) deny that there is any relationship between the two genera on this basis. It is possible that in Britain, where *Pseudobryopsis* does not occur, *Ostreobium* may be a non-specific entity representing phases in the life histories of members of the Bryopsidaceae, though no such link-up has yet been found.

BRYOPSIS Lamouroux

BRYOPSIS Lamouroux (1809), p. 333.

Type species: *Bryopsis pennata* Lamouroux (1809), p. 333.

Thallus of branched coenocytic filaments differentiated into prostrate and rhizoidal system and erect axes variously branched; branches of both unlimited and limited growth; ultimate branches clustered with distichous, spiral or irregular arrangement; branches graded in length giving frond a characteristic conical or feather-like shape; bases of branches constricted and with internal wall thickenings; chloroplasts numerous, disk-shaped or oval with 1–2 pyrenoids; pyrenoids polypyramidal.

Erect plants are gametophytes, dioecious or monoecious; biflagellate gametes anisogamous, formed in terminal branches cut off by basal wall; zygotes producing new gametophytes directly, or giving dwarf filaments producing gametophytes after resting stage or producing stephanospores; dwarf filaments sometimes producing gametes; vegetative reproduction by fragmentation, by development of isolated pieces of cytoplast, or by aplanospore formation.

The life history of *Bryopsis hypnoides* and *B. plumosa* is unusual in that, although there are two different morphological phases, they are almost certainly both diploid. The balance between the two morphological phases, as shown by culture experiments, is different in plants collected from different latitudes, the differences being related to different temperatures. In Anglesey, North Wales, plants of *B. hypnoides* were found to be sexually reproductive from May to September. In culture the zygotes developed into prostrate filamentous germlings a few of which grew directly into adult *B. hypnoides* plants. The majority, however, ceased development after about three months, while a few gave rise to stephanospores. The stephanospores, on germination, gave rise to erect *Bryopsis* plants, which again formed gametes. The dormancy of the filamentous germlings could be broken by changes in culture conditions (Diaz–Piferrer, 1972). This type of life history has already been established for *Bryopsis hypnoides* and *B. plumosa* in other European countries by Neumann (1969b) and Rietema (1969, 1970, 1971). They showed that the filamentous germlings each contain a single large nucleus which divides in the formation of the stephanospores. Rietema (in Neumann, 1974) found that stephanospores from dwarf filaments gave rise to 50 per cent male and 50 per cent female plants in the dioecious species of *Bryopsis*. It was therefore expected that meiosis would occur in the giant nucleus of the dwarf filaments. Neumann (1969), in fact, had assumed that it occurs in the gametangium of the erect *Bryopsis* plant just prior to gamete formation. The evidence of Zinneker (1935) also indicated this site for meiosis. That the gametangium is really the site of meiosis has now been shown by Kermarrec (1980). He has also shown that the large nucleus of the dwarf filaments divides by mitosis and the stephanospores therefore are accessory spores. Diaz–Piferrer (1972) found that sexual reproduction occurred only rarely in plants of *B. hypnoides* in Anglesey and not at all in plants from the tropical waters of Puerto Rico. In Newfoundland, at the northern end of the distribution range for the species, both the erect *Bryopsis* plants and also the filamentous germlings gave rise to gametes (Bartlett & South, 1973). This also fits in with the diploid nature of both morphological phases of the life history. The evidence of Diaz–Piferrer (1972) suggests that vegetative reproduction by a variety of methods may predominate in the field.

Kornmann & Sahling (1976) have recently examined the validity of *Bryopsis*

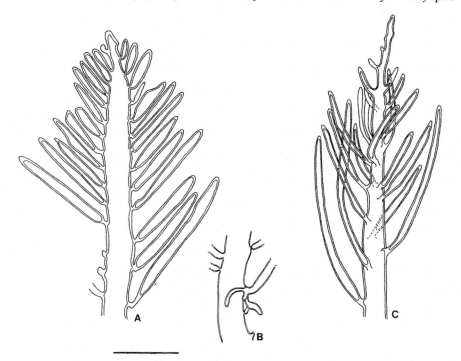

Fig. 57   *Bryopsis spp.*
A. *Bryopsis plumosa*. Apex region. (Orkney). B. *B. plumosa*. Basal branch with rhizoidal processes grasping axis. (Orkney). C. *B. hypnoides*. Apex. Bar= 500 µm.

lyngbyei Hornemann. They found that the pinnules are arranged distichously and the life history includes a filamentous sporophyte stage which produces only stephanokont zoids. 'Crosses with samples of *B. plumosa* from European coasts were unsuccessful; thus *Bryopsis lyngbyei* is evidently a separate species.' If *B. lyngbyei* occurs in the British Isles, it is completely confused with *B. plumosa*.

A number of workers, including the present author, have had great difficulty in separating *B. plumosa* from *B. hypnoides* as they overlap morphologically. The whole question of their separation needs to be investigated further.

KEY TO SPECIES

Ultimate branches arranged distichously . . . . *B. plumosa*
Ultimate branches arranged spirally or irregularly . . . . *B. hypnoides*

Bryopsis hypnoides Lamouroux (1809), p. 135.   Fig. 57B

Type: CN. Cette, Mediterranean.

Bryopsis monoica Funk (1927), p. 332.

Thallus heteromorphic; erect plants light to dark green; branched filaments 10 cm or more high arising from prostrate rhizoidal branch system; erect filaments with one to several degrees of branching; axes and main branches of indeterminate growth, up to 500 μm diameter, arranged spirally or irregularly; ultimate branches in clusters, of determinate growth, with swollen bases, up to 60 μm diameter, lengths graded to give frond conical shape; sometimes thick rhizoidal outgrowths at bases of main and secondary branches clasping axes; chloroplasts oval, 7–10 μm long with single pyrenoid; microscopic branched filaments each with single large nucleus also occur.

Reproduction mainly vegetative by fragmentation, by regeneration of pinnules and of isolated pieces of cytoplast, also by aplanospores; sexual reproduction by biflagellate gametes formed in gametangia from laterals of determinate growth; microgametes at distal end of gametangium, macrogametes at proximal end, though positions may be reversed; gametes sometimes in separate gametangia; plants usually monoecious; asexual reproduction by stephanospores formed in dwarf filaments.

Chromosome no.: $2n = 12$ Kermarrec (1980).

Found growing in well-illuminated and usually clear rock pools in the mid- to lower littoral region and in the sublittoral, in sheltered and also exposed parts of the shore; in sheltered areas often attached to stones and shells and may form large tufts more or less free-floating; sometimes found on sand-covered rocks.

Distributed all round the coast of the British Isles in suitable habitats.
Scandinavia south to the Atlantic coast of France; also recorded from the Mediterranean.
The Atlantic coast of North America from Newfoundland to southern Massachusetts, Florida, Bermuda, Mexico, Puerto Rico: Pacific coast of North America from British Columbia to southern California, Panama; Sri Lanka, Japan.

Plants can be found at all times of the year, but most conspicuously in the spring.

Records of sexual reproduction have been between May and September; dwarf filaments and stephanospore formation have not been recorded in the field. For life history see comments on the genus (p. 180).

The arrangement of branches on the axes is very variable in this species and may at times appear almost distichous, the species then showing similarities to *Bryopsis plumosa*, though the plants retain their overall conical shape. Sometimes rhizoidal outgrowths appear at the bases of both primary and secondary branches; these curl round and clasp the axes. The particular conditions under which they are formed have not been described.

Bryopsis monoica Funk was separated on its monoecious condition; this condition is now known to occur in *B. hypnoides* (Diaz-Piferrer, 1972). According to Funk (1927) *Bryopsis pulvinata* Oltmanns is synonymous with *B. monoica* Funk.

It has been shown by Tatewaki & Nagata (1970) and by Diaz-Piferrer (1972) that extruded protoplasts can synthesise new walls and can develop into normal plants.

Bryopsis plumosa (Hudson) C. Agardh (1832), p. 448.   Fig. 57A, B; Pl. 8

Type: missing. Exmouth, Devon.

Ulva plumosa Hudson, (1778), p. 571.

Thallus heteromorphic; erect fronds of light to dark green branched filaments 5–10 cm or more high, arising from prostrate rhizoidal-attaching system; erect filaments and primary branches of indeterminate growth with terminal clusters of branches of determinate growth arranged distichously, lengths graded to form flat triangular shape; lower branches often with rhizoidal processes grasping axes; main axes 500–1500 µm diameter, branches 50–250 µm diameter; chloroplasts disk-shaped or spindle-shaped, with 1–2 pyrenoids; microscopic branched filaments with single large nucleus.

Reproduction by biflagellate gametes, anisogamous; plants usually dioecious, occasionally monoecious; gametangia formed from determinate lateral branches cut off by double septum; microgametangia light brown, macrogametangia dark brown; asexual reproduction by stephanospores formed in microscopic filaments; vegetative reproduction by fragmentation and regeneration of lateral branches, of isolated pieces of cytoplast and by aplanospores.

Chromosome no.: 2n = 8 Zinnecker (1935).
                 2n = 14 Kermarrec (1980).

Found in clear, rocky, often deep pools, but also in sandy pools in the mid- to lower littoral region, commonly under overhanging ledges, and in the sublittoral down to 8–15 m; also on stones and shells in very sheltered places.

Generally distributed round the whole coast of the British Isles in suitable habitats. Scandinavia south to the Mediterranean, Baltic Sea.

Canary Islands, Azores, Sénégal, Ghana, Mauritania, South West Africa; Japan, southern Australia; Maritime Provinces of Canada south to Connecticut on the Atlantic coast of North America, Antigua; southern British Columbia south to central California on the Pacific coast of North America.

The seasonal behaviour of *Bryopsis plumosa* is not very well established in the British Isles. Edwards (1975) regards it as a summer and autumn annual; Knight & Parke (1931) regarded it as a perennial, small and scattered plants being found in the winter in deep pools in the Isle of Man.

For the life history see comments on the genus (p. 180).

Womersley (1956) records plants as distichous, but also sometimes with ramuli on all sides of the axis. This is a general experience and morphologically it is sometimes extremely difficult to separate this species from *Bryopsis hypnoides*. The life history and environmental modifications are also similar in the two species. Axes of *B. plumosa* are sometimes bare or almost bare of branches and this form has been described as var. *simplex* Holm. & Batt. and var. *nuda* Holm. (see Batters, 1902).

DERBESIA Solier

DERBESIA Solier (1847), p. 157.

Type species: *Derbesia marina* (Lyngbye) Solier (1847), p. 158.

Halicystis Areschoug (1850), p. 447.

Sporophyte forming soft tufts of erect branched filaments arising from prostrate system of branched filaments with lobed haptera; degree of branching variable; septa delimiting a single small cell sometimes formed at base of branch; chloroplasts numerous, spindle-shaped or discoid, with or without pyrenoids; gametophyte an ovoid or spherical bladder attached by rhizoids penetrating tissues of calcareous substrate.

Asexual reproduction by uninucleate, multiflagellate zoospores formed in ovoid sporangia lateral to erect filaments; flagella in flattened ring at anterior end of zoospore; sexual reproduction by macro- and microgametes, biflagellate, formed in patches in upper end of gametophyte bladder; plants dioecious; gametes released through one or more pores which later close; several series of gametes may be formed.

KEY TO SPECIES

1 Plant filamentous 2
  Plant vesiculate 3
2 Chloroplasts discoid, lacking pyrenoids . . . . *D. marina*
  Chloroplasts spindle-shaped with one or more pyrenoids . . . . *D. tenuissima*
3 Vesicles with discoid chloroplasts lacking pyrenoids; gametangia lateral,
    irregularly band-shaped with several pores . . . . *D. marina* (= *Halicystis ovalis*)
  Vesicles with spindle-shaped chloroplasts with 1–2 pyrenoids; gametangia
    terminal, more or less circular with lobed edges and single pore . . . .
    . . . . *D tenuissima* (= *Halicystis parvula*)

Derbesia marina (Lyngbye) Solier (1847), p. 158.   Fig. 58

Holotype: C (a specimen collected by Hornemann for Lyngbye. P. S. Dixon, pers. comm., reported that there is no specimen of Lyngbye's in C, but as Hornemann is

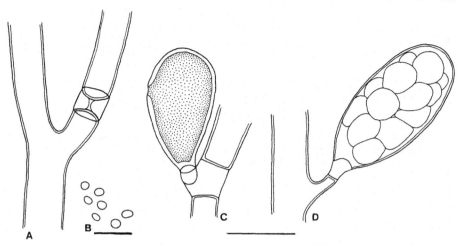

Fig. 58   *Derbesia marina*
A. Portion of erect filament showing cross wall at base of branch. (Chapmans Pool, Dorset). B. Chloroplasts. C. Young sporangium. (Chapmans Pool, Dorset). Bar=. D. Mature sporangium with enclosed stephanospores. (Dredged, Isle of Man; Aug. 1952; S. Lodge). Bar= 50 μm (A, C, D), 10 μm (B).

known to have collected for Lyngbye it is reasonable to assume that the plant collected by him is in fact the type specimen from the district in the Faroes cited by Lyngbye). Quivig, Faroes.

Vaucheria marina Lyngbye (1819), p. 79, tab. 28, fig.1.
Halicystis ovalis (Lyngbye) Areschoug (1850a), p. 447 (reprint p. 221).
Gastridium ovale Lyngbye (1819), p. 72.

Sporophyte forming soft bright green silky tufts of filaments up to 1–3 (–4) cm long; filaments laterally or pseudodichotomously branched, arising from a basal creeping system penetrating the substrate; filaments lacking cross walls except at bases of branches where double walls sometimes cut off short cells; filaments 15–20 (–70) μm diameter; chloroplasts discoid, lenticular, 2–3 μm diameter, lacking pyrenoids; gametophyte (= *Halicystis ovalis* (Lyngbye) Areschoug) vesiculate, clavate or spherical, 10–15 mm diameter, attached by slender rhizoids penetrating substrate; chloroplast discoid, lacking pyrenoids.

Reproduction by stephanospores formed in pyriform or clavate sporangia, 100–190 (–300) × 60–80 (–150) μm, attached laterally to filaments by short stalk subtended by double-walled narrow cell; zoospores 16–32 (–48) per sporangium; individual sporangia can revert to vegetative state; sexual reproduction by biflagellate gametes, anisogamous, formed in single fertile patches of gametophyte as an irregular transverse band, with several pores; vegetative reproduction by regeneration from rhizoidal system and by fragmentation.

Chromosome no.: n = 8, 2n = 16 Neumann (1969).

Plants occur in the sublittoral region from 2–20 (–24) m depth attached to sponges, shells and other algae on both sheltered and exposed coasts, but usually where there is rapid water flow; vesicular phase grows directly on *Lithothamnion* (=coralline)-covered rock surface and on other organisms; in Japan it has been reported to occur in the lower littoral.

Generally distributed round the coasts of the British Isles and may be more common than is indicated by the actual records.

Shetland, Argyllshire, Bute, Isle of Man, Isle of Wight, Hampshire, Dorset, Devon, Co. Dublin, Co. Waterford, Co. Mayo.

Norwegian Arctic Sea south to the Mediterranean and the Canary Islands.

Greenland, Faroes, eastern arctic Canada south to Massachuesetts and North Carolina, west coast of North America from Alaska to Mexico; Bahamas; Brazil; Japan, South Australia.

On the eastern coast of North America the filamentous stage is found much more widely than the vesicular. The '*Halicystis*' phase has so far not been found in Newfoundland and only recently in New England (Sears & Wilce, 1970; Mathieson & Burns, 1970).

Plants can be found at all times of the year; they can perennate as endozoic and endophytic filaments that regenerate. Sporophytes have been reported as fertile in the British Isles during May, June and July, gametophytes in August, but this is probably not the complete picture. For New England, Sears & Wilce (1970) found that sporophytes were fertile in the field only at temperatures below 13°C and most commonly during the winter months at 0–5°C. Some reproductive plants were observed during August, but only in a single collection made at a depth where the temperature was 12°C as opposed to surface temperature of 18°C. In culture, plants produced sporangia at 15–20° but not at 5–10°C.

The life history is an alternation between filamentous sporophyte and vesiculate gametophyte, meiosis occurring in the formation of zoospores; zoospores can sometimes germinate directly to reform haploid filamentous plants (Kornmann, 1966; Neumann, 1969; Sears & Wilce, 1970); undivided sporangia can sometimes form gametophytes or revert to the vegetative state (Sears & Wilce, 1970; Neumann, 1969).

Kornmann (1970) described a mutant of *Derbesia marina* which Neumann (1969) showed to be haploid. It had spindle-shaped chloroplasts, larger than normal, 3–4 μm in diameter; branches of the filaments were also abnormal. Sporangia were formed but no meiosis occurred before zoospore formation.

Derbesia tenuissima (De Notaris) Crouan frat. (1867), p. 133.

Type: RO? Capraria Island, Mediterranean (see Kobara & Chihara, 1981).

Bryopsis tenuissima Maris & De Notaris (1839), p. 259.
Halicystis parvula Schmidt (1819), p. 72.

Sporophyte forming compact green tufts, 1–4 (–8) cm high, attached by creeping filaments penetrating substrate; erect filaments (25–) 30–50 (–70) μm broad, sparsely branched, branching dichotomous or irregular; chloroplast spindle-shaped or irregular,

8–10 μm long with 1–2 pyrenoids. Gametophyte (= *Halicystis parvula* Schmidt in Murrey) spherical or ovoid, 10–15 μm diameter, attached by short pedicel, more or less swollen at tip, cavity continuous with vesicle, pedicel penetrating calcareous substrate; chloroplasts spindle-shaped with 1–2 pyrenoids.

Reproduction by stephanospores formed in ovate, globular or pyriform sporangia, somewhat flattened at distal end, in communication with filament or cut off by plug about 12 μm thick; sporangia (90–) 150–250 × 115–230 (–300) μm; sexual reproduction by biflagellate gametes, anisogamous, formed in gametangia at apex of gametophyte vesicle, forming more or less circular patches with lobed edges and usually a single pore; female gametes 10 × 14 μm, male 3 × 8 μm; vegetative reproduction by proliferation of prostrate filaments and by isolation and regeneration of cytoplast material, by fragmentation and regeneration of erect filaments.

Chromosome no.: n = 8 Neumann (1969).

The filamentous stage is found in pools at extreme low water, and in the sublittoral region, growing on rocks, other plants and on animals; vesiculate stage found growing on calcareous or '*Lithothamnion*' (=coralline)-covered rocks free of silt, in the sublittoral.

There is no definite record of the vesiculate stage in the British Isles. The filamentous stage is very rare and apparently confined to the southern counties.
Dorset, Devon, Isle of Wight; the last is the only recent record.
Atlantic coast of France, Mediterranean, Canary Islands.
Japan.

There is practically no information on the seasonal behaviour of this species in the British Isles. Only the erect filamentous sporophyte has been found here and the fruiting period for the few records is for the summer months. Hamel (1931) records it as present all the year round in the Mediterranean with a fruiting period between February and September.

No information on form variation.

There is evidence for an endogenous rhythm controlling gamete formation in this species in the vesicular stage (Page & Sweeny, 1968; Page & Kingsbury, 1968) with a basic period of 4–5 days.

OSTREOBIUM Bornet & Flahault

OSTREOBIUM Bornet & Flahault (1889), p. 161.

Type species: *Ostreobium quekettii* Bornet & Flahault (1889), p. 161, pl. IX, figs 5–8.

Thallus endozoic, of multinucleate, non-septate, irregularly branched filaments with or without irregular swellings at intervals; filament walls sometimes thick and/or striate; chloroplasts small, elongate spindle-shaped, spherical or tetrahedral.
Reproduction by quadriflagellate zoospores and aplanospores.
The status and systematic position of the genus *Ostreobium* are uncertain. Mayhoub

(1974 *a*, *b*) found that filaments very similar to *Ostreobium* occur as 'protonema' in the life history of *Pseudobryopsis myura*. *Pseudobryopsis* is not known to occur in the British Isles, but it is possible that the protonemal stage might have a much wider distribution range than other parts of the life history. If *O. quekettii* and *P. myura* prove to be conspecific then the present position of *Ostreobium* in the Bryopsidaceae might be justified though *Ostreobium* differs from all the other genera of the family in having quadriflagellate zoospores and not stephanospores. Kornmann & Sahling (1980) on the basis of their culture work did not think that the 'prothallus' described by Mayhoub (1974 *a*, *b*) as part of the life history of *Pseudobryopsis myura* was identical with European *Ostreobium*. They place the genus *Ostreobium* in the Codiales but do not commit themselves further.

Only one species in Britain.

Ostreobium quekettii Bornet & Flahault (1889), p. 161, pl. IX, figs 5–8.

Type: PC. Croisic, Brest, and Normandy, France.

Pseudobryopsis myura (Agardh) Berthold. (See Mayhoub, 1974 *a, b* for a discussion of this synonym).

Thallus of highly branched tubular coenocytic filaments 2–5 μm diameter, constricted between irregular swellings up to 40 μm diameter, fine ends of filaments about 2 μm diameter; contents of filaments dense; chloroplasts small, elongate, spindle-shaped, lacking pyrenoids.
    Reproduction by quadriflagellate zoospores, 9–12 μm long, formed in irregularly swollen ends of branches, cut off by membrane; zoosporangia with shorter or longer hyaline exit tubes; reproduction also by aplanospores formed at filament ends.

Chromosome no.: not known.

Occurs in shells of molluscs and calcareous tubes of worms, frequently together with other shell-boring species, in the mid–lower littoral and the sublittoral region. Recorded hosts: *Pomatoceros* tubes, *Spirorbis* spp., *Littorina littoralis*, *Patella vulgaris*, *Nucella lapillus* and old oyster shells and mussel shells.

Widely distributed round the coasts of the British Isles and probably more common than is indicated by the records.
    Lewis, Harris, Argyllshire, East Lothian, Berwickshire, Isle of Man, Lundy Island, Dorset, Devon, Co. Clare, Co. Wexford.
    Found from Scandinavia south to the Mediterranean and the Canary Islands, Baltic Sea.
    Eastern North America from Ellesmere Island south to the Caribbean, Greenland, Iceland, Faroes; Sri Lanka, Easter Island, Maldive Islands, Samoa, Australia, Caroline Island, Marshall Islands.
    The species is said to occur world-wide, but Kornmann & Sahling (1980) think that further evidence is required before all these records can be accepted as referring to plants identical with the type species.

The occurrence of the species appears to be sporadic, but it can occur at any time of the year; it is recorded as present in *Spirorbis* in November, January, March and July (Wilkinson, 1969). Nothing seems to be known about the reproductive period in this country, but it has been recorded as fertile in summer on the west coast of Sweden (Kylin, 1949).

Bornet & Flahault (1899) described the lateral branches as fusing together to form a net-like structure which they illustrated. Kornmann & Sahling (1980) found no net-like structure in their cultures of *Ostreobium*. However they do not suggest that they have seen the type material.

It might be that plants like those of this species have originated as part of the life histories of several algae and become independent in the way suggested by Feldmann (1952) and Bernatowitz (1958).

CODIACEAE Feldmann

CODIACEAE Feldmann (1954), p. 97.

Thallus uncalcified or calcified, of branched coenocytic filaments interwoven to form solid spongy structure attached by prostrate system of rhizoidal filaments, sometimes forming an attachment disc; thallus shape various, cushion-like, stalked leaf-like, fan-shaped with or without lobed segments, bottle brush, globular, erect terete dichotomously branched, dendroid; internally differentiated, longitudinally running narrow branched filaments forming medulla, ultimate branches forming close palisade cortex of swollen utricles with laterally attached hairs; utricles with or without mucron; chloroplasts numerous, embedded in cytoplasmic lining of filaments, disc-shaped without pyrenoids.
    Sexual reproduction by anisogamous biflagellate gametes formed in gametangia as lateral branches of utricles; plants monoecious or dioecious; reproduction also vegetative.

CODIUM Stackhouse

CODIUM Stackhouse (1797), p. XXIV.

Type species: *Codium tomentosum* Stackhouse (1797), p. XXIV. Stackhouse's illustration of *Fucus tomentosus* (1795, pl. 7) is undoubtedly *C. vermilara*.

Lamarckia Olivi (1792), p. 258 pl. VII.
Myrsidium Rafinesque (1810), p. 98
Spongodium Lamouroux (1813), p. 72.
Agardhia Cabrera (1823), p. 99.

Thallus of branched coenocytic filaments, sometimes persisting as undifferentiated filaments in loose tangles or turfs, more usually organised into a definite form, either globular or erect with dichotomously branched more or less terete axis, or forming a spongy flat or convoluted carpet over substrate; mature thallus spongy, of interwoven branched filaments forming central medulla and cortex of closely packed, swollen utricles with annular thickenings at bases; utricles variously shaped, sometimes branched, frequently bearing colourless hairs in positions which are characteristic for each species; utricle ends rounded or flat, sometimes bearing longer or shorter mucrons.

Biflagellate anisogametes formed in gametangia developed as lateral branches of the utricles cut off by annular thickenings; plants usually dioecious; macrogametes sometimes parthenogenetic.

The life history is known for few species: for these the plants are diploid and the only haploid phase is represented by the gametes.

KEY TO SPECIES

1 Frond erect, terete, dichotomous or irregularly branched, arising from a
    broad or narrow basal disk with or without a spreading prostrate
    system. 2
    Frond prostrate, forming a velvety layer over substrate . . . . *C. adhaerens*
    Frond globular, attached by a portion of lower surface. . . . . *C. bursa*
2 Some or all utricles mucronate 3
    Utricles never mucronate. 4
3 Frond terete throughout: utricles with rather blunt mucrons, less than
    15 µm long . . . . *C. fragile* subsp. *atlanticum*
    Frond more or less flattened below dichotomy: utricles with very pointed
    mucrons up to 68 µm long . . . . *C. fragile* subsp. *tomentosoides*

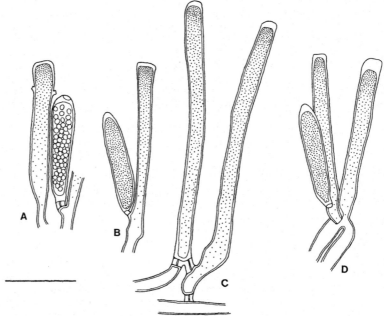

Fig. 59   *Codium adhaerens*
    A, B, D. Utricles with gametangia. (Salcombe, Devon; Oct. 1953; C. Bliding). C. Two
    utricles. (Portscatho; Oct. 1977; S. Hepton). Bar= 200 µm.

4   Frond more or less strictly dichotomously branched; utricles with broad
    base and broad rounded apex with narrower region between; hairs or
    hair scars on constriction below apical swelling . . . . *C. tomentosum*
    Frond irregularly branched, branches sometimes appearing proliferous;
    utricles broadening from base to rounded or rather flat apex; hairs or
    hair scars attached just below rim of broad apex . . . . *C. vermilara*

Codium adhaerens C. A. Agardh (1822), p. 457.   Fig. 59

Type: LD (No. 15465 in Agardh herb.; along with the specimen there is a letter from
Cabrera to Agardh giving information on the species; Dellow, 1952).

Agardhia adhaerens (C. Agardh) Cabrera (1823), p. 99.

Thallus forming dark green felt, irregular in shape, closely adherent to rock surface, 5–
10 mm thick, smooth or corrugate, lobed, occasionally partially raised from substrate;
texture gelatinous, firm; utricles cylindrical with flat or rounded ends, 40–100 μm
broad, 500–800 μm long; hairs attached towards apex of utricle.

    Reproduction by biflagellate gametes; gametangia, 60–75 μm broad, formed on
lower part of utricle; plants dioecious or occasionally monoecious.

    Chromosome no.: not known.

    Covers rock surface up to 1 m² or more in the lower littoral and sublittoral fringe in
both sheltered and exposed sites under low light conditions particularly where there is
a strong flow of water. Plants of this species have been found on a sheltered north-
facing wall in the upper littoral in Cornwall (Price, Hepton & Honey, 1980).

    Probably more common than is indicated by the actual records which are scattered
round the coast of the British Isles: Invernessshire, Argyll, Dorset, Devon, Cornwall,
Channel Islands, Co. Antrim, Co. Donegal.
    British Isles south to Portugal and the Mediterranean.
    Brazil, West Indies, west coast of Africa, Mauritius, Red Sea, Indian Ocean, Sri
Lanka, Japan, Australia, New Zealand, Friendly Islands, Hawaii, Dutch East Indies,
Bermuda.
    Early distribution records suggest that the species formerly had a wider geographical
range than now. This is probably due to the confusion between *C. adhaerens* and *C.
difforme*. These two species differ in the thickness of the frond, in the size of the
gametangia and in texture (Vouk, 1936). According to this author, all of the records
from the Adriatic Sea should be referred to *C. difforme*. The plants originally
identified as *C. adhaerens* on the Pacific coast of North America are now referred to *C.
setchellii* Gardner.

    Plants appear to be present all the year round, but according to Cabrera (Letter in
LD) '*Codium adhaerens* flourishes for a certain length of time and then dies away to
reappear the following year.' Experiments by Dellow (1952) support this. The plants
are fertile through summer, but especially towards October, rarely fruiting in some
localities.
    The life history is not known.

    Two varieties of *C. adhaerens* have been described from New Zealand by Dellow

(1952) depending on the smoothness or corrugation of the thallus surface. It seems, however, that this can be a variable character at a single site.

Codium bursa (Olivi) C. Agardh (1821), p. 457.

Type locality: Adriatic.

Lamarckia bursa Olivi (1792), p. 258

Thallus a flattened spongy sphere with velvety surface, up to 30 cm across, attached by interwoven filaments; interior with loosely packed branched filaments leading to cortex of tightly woven filaments ending in clavate utricles; thallus becoming hollow or water filled with age; utricles 250–550 μm broad, up to 4.5 mm long, slightly constricted below rounded apex (Schmidt, 1923); hairs present, concentrated at apex of utricle.

Reproduction by gametes formed in spindle-shaped or ovoid gametangia, one or several to each utricle.

Chromosome no.: not known.

Attached to shells and rock surface in the sublittoral down to 10 m or more; often occurring in small groups, probably in rather sheltered positions.

A rare British species. Many of the existing records refer to drift specimens, but it has been recorded attached; a recent record comes from Mulroy Bay, Co. Donegal where it was found growing attached to rock in about 10 m of water. The specimen is in the Ulster Museum and was determined by Mr O. Morton.

Sussex, Devon, Cornwall, Co. Donegal, Channel Isles, Co. Antrim.

European Atlantic coast from Belgium and the British Isles to Portugal, Mediterranean, Canary Isles, Adriatic Sea.

Not recorded outside Europe, but the Australian *Codium mamillosum* Harvey is very similar.

There is little information on the seasonal behaviour of this species though it is known to be present during the summer months. Nothing is known about its reproductive period in the British Isles and its life history is not known.

Codium fragile (Suringar) Hariot (1890), p. 32.

Type: L (No. 963 according to Scagel, 1966, p. 119; it has 'consistently and sharply pointed spines at the apex of the utricles'). Japan.

Acanthocodium fragile Suringar (1870), p. 23, Tab. VIII.

Thallus erect, up to 30–40 cm or more high, usually dichotomously, sometimes irregularly branched, terete or somewhat flattened below nodes, dark green, attached singly or with several fronds arising from a broad or narrow spongy disc, sometimes merging into a spreading prostrate filamentous system; utricles cylindrical or club-shaped, 100–400 μm broad, 330–1500 μm long with shorter or longer mucron at rounded outer end where the wall is often thickened; utricles with 1–2 colourless hairs in upper part.

Gametangia ovoid, clavate or spindle-shaped, attached to utricle by protuberance towards upper end; plants dioecious; gametes sometimes parthenogenetic.

Growing on rocky shores from mid-littoral to sublittoral.

Grows all round the coast of the British Isles on suitable substrates, as one or other of two species.

Norway, Denmark, Netherlands.

Japan, Australia, New Zealand, South Africa, Indian Ocean; American Pacific coast from Alaska to Cape Horn.

Codium fragile (Sur.) Hariot was first recorded in the British Isles, on the west coast of Scotland, as *C. elongatum* Ag. by J. Agardh (1886; Holmes & Batters, 1890), although herbarium specimens of plants collected on the Ayrshire coast in 1839, kept in the Dublin National Museum, show that the species has been here much longer (Cotton, 1912). In 1897 Batters received a specimen from Kilkee, Co. Clare, collected by E. George, which he identified as *C. elongatum* Ag. and listed it in his 'Catalogue' (1902), giving Kilkee as the only locality (Cotton, 1912). Cotton (loc. cit.) doubted this identification, but was certain that the plant was the same as plants he had collected on Clare Island in 1910 which he identified as *C. mucronatum* var. *atlanticum* var. nov. Silva (1955) described two subspecies of *C. fragile* (Sur.) Hariot for Britain, one of which (*C. fragile* subsp. *atlanticum*) was conspecific with Cotton's plant mentioned above. The second subspecies was *C. fragile* subsp. *tomentosoides*.

The species *Codium fragile* has an almost world-wide distribution, the centre of which is given by Silva (1955) as the Pacific and Subantarctic Oceans. It consists of a large complex of populations some of which, according to Silva (loc. cit.) are morphologically homogeneous, others heterogeneous. At the centre of distribution, the populations may show overlap or intermediate characters, but they are much more distinct the further from the centre they are found.

In Britain the two subspecies of *C. fragile* described by Silva (1955) show little in common beyond their erect dichotomous form and the presence of a mucron on the utricle. They differ considerably in the following characters: thallus thickness, presence/absence of a flattening of the thallus below a dichotomy, length of the mucron, structure of the mucron, life history as far as it is known, shore position, distribution in the British Isles, wall chemistry and fine structure.

C. fragile subspecies *atlanticum* belongs to the northern part of the British Isles, growing only as far south as Co. Cork in Ireland, Anglesey in the Irish Sea and Northumberland on the North Sea coast; subspecies *tomentosoides* belongs to the south coast, but has spread northwards: its present northerly limits appear to be the west and north coast of Ireland, and the southwest coast of Scotland. It does not appear to have penetrated into the Irish Sea. The subsp. *tomentosoides* was first found in Britain in 1939 at Start Point in Devon and its progress along the south coast and northwards can be followed by records in the literature since then.

When the two subspecies *atlanticum* and *tomentosoides* were found together in any quantity on the same shore, as was the case in Co. Clare in 1958 (pers. obs.), subspecies *atlanticum* occupied a belt in the *Fucus spiralis*/ upper *F. vesiculosus* zone and subspecies *tomentosoides* grew in the *Fucus serratus* zone. No real evidence was found of plants with characters intermediate between the two subspecies. This could perhaps be explained by the fact that parthenogenesis occurs in subspecies *tomentosoides* (Dangeard & Parriaud, 1956; Delépine, 1959; Feldmann, 1956), involving the production of large numbers of macrogametes which are liberated to develop directly on the shore. The efficiency of this mode of reproduction may account for the rapid spread of this subspecies during the last two decades.

The two subspecies also differ considerably in the fine structure of their chloroplasts and primary wall layers and in the chemistry of the primary walls as indicated by staining reactions (Gibby, 1971). These differences argue that the two taxa are not so closely related as their present taxonomic situation implies, certainly at this end of the distribution range of *C. fragile*. There is no evidence of gene-flow between the subspecies in Britain. It was suggested by de Valera (1939) that *C. fragile* subsp. *atlanticum* should be afforded specific rank and Gibby (1971) suggested the combination *Codium atlanticum* (Cotton) de Valera. In view of the recent arrival in this country of subspecies *tomentosoides* and the fact that its source of origin is not known, perhaps further study of its overall distribution and relationships is required before elevation is contemplated for this subspecies. No proposal has therefore been made for raising subspecies *tomentosoides* to full specific status and subspecies *atlanticum* is also retained at this level, though almost certainly full specific status will eventually be justified for both.

KEY TO SUBSPECIES

Mucron <20 μm long, plants terete, not flattened at nodes . . . .
. . . . *C. fragile* subsp. *atlanticum*
Mucron long and pointed, plants flattened at nodes . . . *C. fragile* subsp. *tomentosoides*

Codium fragile (Suringar) Hariot subsp. *atlanticum* (Cotton) Silva (1955), p. 565.
Fig. 61A

Holotype: BM–K (specimen collected by Cotton in June 1910). Roonah Point, Co. Mayo, Ireland.

Codium mucronatum var. *atlanticum* Cotton (1912), p. 114.
Codium tomentosum var. *atlanticum* (Cotton) Newton (1931), p. 106.

Thallus light green, robust, terete, not flattened below dichotomies, 10–25 cm high, rarely more, attached by a stout base; utricles large, more or less cylindrical or somewhat clavate, 800–1200 μm long × (130–) 250–300 μm broad, apex rounded, thin or thick-walled, sometimes flattened and indented with short stout mucron < 20 μm long, sometimes absent in all but young utricles; hairs 1–2 per utricle attached about 300 μm below apex.

Gametangia formed on short stalks arising near middle of utricle, 1–2 per utricle, ovate, clavate or fusiform, 80–120 μm broad × 250–400 μm long; plants dioecious.

Chromosome no.: not known.

Fig. 61   *Codium spp.*
           A. *C. fragile* subsp. *atlanticum*. Utricle. B. *C. fragile* subsp. *tomentosoides*. Utricles.
           C. *C. tomentosum*. Utricle with gametangia. D. *C. vermilara*. Utricle. g= gametangia,
           hs = hair scars. Bar= 100 μm.

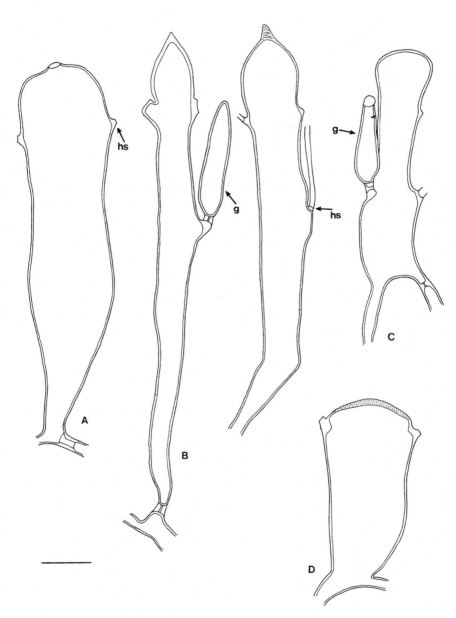

Grows in clear pools in any shore position other than the very top of the littoral zone; on the lower shore it grows on rock surfaces over which water drains slowly; also in the shallow sublittoral.

It occurs in Shetland and down both the eastern and western coasts of Scotland; south to Anglesey in the Irish Sea and to Northumberland in the North Sea; it is found along the west and north coasts and also in places along the east and south coasts of Ireland; a record from the south coast of England from Swanage, Dorset, is out of the range for the species. Silva (1955) queried the correctness of the location. The record for Dorset (Burrows, 1964) was an uncorrected typing error. In spite of continued search since 1959 no specimen of this subspecies has been found in Dorset.

Norway.

Not found outside Europe.

The plants appear to be perennial or pseudoperennial. Parkes (1975) working in Ireland found a seasonal growth rhythm for this subspecies. The erect fronds disappeared in December or early January leaving a mossy growth of tangled unorganised threads. In April or early May new erect fronds arose which grew rapidly, reaching peak development in late summer and early autumn, deteriorating gradually as winter set in. Cotton (1912) for Clare Island found young plants in February with maximum development in summer; he thought that they disappeared in winter though he remarked that a few sporeling could be found all the year round. Fertile plants have been recorded in summer (Knight & Parke, 1931 and pers. observ.) and in October and December (Parkes, 1975).

The plants can exist as tangles of unorganised branched filaments or as a short turf on the rock surfaces which are very difficult to identify until the characteristic erect fronds are found arising from them.

Codium fragile (Suringar) Hariot subsp. *tomentosoides* (Van Goor) Silva (1955), p. 567.   Fig. 61B, C; Pl. 9

Holotype: missing. (Huisduinen, Netherlands, 1900, collected by Mrs J. L. Redeke-Hoek).

Neotype: AMD (designated by Silva (1955): a specimen from Helder, Netherlands, annotated by Van Goor).

Codium mucronatum J. Agardh var. *tomentosoides* Van Goor (1923), p. 134.

Thallus dark green, soft, up to 1 m or more long, terete, flattened below dichotomies; utricles usually more or less clavate with constriction at or just below middle, sometimes a second constriction lower down; utricles (105–) 165–325 (–400) µm max. diameter, 550–1050 µm long; apices rounded with very pointed mucron up to 68 µm long; wall of mucron with fine concentric striations; 1–2 hairs per utricle, attached about 200 µm below apex.

Gametangia borne on short protuberances near middle of utricle, 1–2 per utricle, ovoid, oblong or fusiform, 72–92 µm diameter, 260–330 µm long.

Reproduction is by parthenogenetic macrogametes.

Chromosome no.: $2n = 40$ for *C. elongatum* Schussnig, 1932 (see Godward, 1966).

In pools and on rock surfaces over the mid-lower littoral region, also in the sublittoral down to 1–2 m, often occurring in thick sheets both in the lower littoral and in the sublittoral.

Scilly Isles, Channel Islands, south coast of England, south and west coast of Ireland, west coast of Scotland to Argyllshire and possibly further north.
For detailed distribution in Ireland see Parkes (1975).
Scandinavia south to the Atlantic coast of Spain.
Atlantic coast of North America; Japan (?).

The total distribution of this subspecies is difficult to extract from the literature, but according to Silva (1955) its place of origin is probably Japan. It was first found in Europe in the Netherlands in 1900 (Van Goor, 1923) and later records indicate that it has spread both north and south along the Atlantic coast of Europe. The first British record was from Devon in 1939 and since then it has spread to the Scilly Isles, Channel Islands and along the south coast of England, and the south and west coasts of Ireland and northwards along the west coast of Scotland to the Orkney Islands. It has also been found in Berwickshire. It does not appear to have penetrated the central area of the Irish Sea.

The plants are present all the year round, but little appears to have been recorded concerning the seasonal pattern of growth. Plants have been recorded as producing gametangia from August to December.
The full life history is not known.

There is considerable variation in the size of fruiting plants, some reaching more than 1 m in length. There is variation also in the breadth of the frond and in the degree of flattening below the dichotomies, some fronds having an external appearance similar to that of *Codium elongatum*.
Recently Hepton (see Price, Hepton & Honey, 1980) has found *Codium fragile* subsp. *tomentosoides* growing as a cushion on rocks near the top of a shore in Cornwall, with young utricles arranged in a palisade, only some of them showing the characteristic mucron.

Codium tomentosum Stackhouse (1795), p. XXLV.   Fig. 60, 61D

Holotype: missing (on 'the Devonshire and Cornwall coasts; on the long rock between Marazion and Penzance plentiful Hon. Mr Wenman.')

Neotype: LINN (selected by Silva, 1955; collected, presumably, by Stackhouse). Acton Castle, Cornwall.

Fucus tomentosus Hudson (1778), p. 584, excl. syn.

Thallus of one to several erect, dichotomously branched fronds arising from a small spongy basal disc or from a more extensive spongy layer of tissue closely adhering to the substrate; erect thalli up to 20 cm high, branches terete, sometimes slightly flattened below dichotomy, terminal segments short; utricles clavate to pyriform 600–800 μm long, 100–200 μm broad, apices rounded; apical wall of utricle slightly to moderately thickened, 5–15 (–50) μm, lamellate: hairs or hair scars common or numerous, borne on waist below broadest part of utricle, 60–110 μm below apex; medullary filaments 18–45 μm diameter.

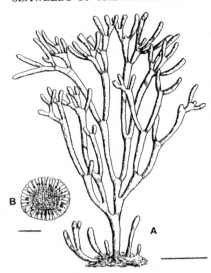

Fig. 60   *Codium tomentosum*
A. Habit of plant. Bar= 30 mm. B. Tranverse section of thallus. (after Newton, 1931).
Bar= 750 μm.

Gametangia oblong to fusiform, (52–) 60–80 (–96) μm diameter, 1–4 per utricle, each borne on protuberance ½–⅔ distance below apex.

Chromosome no.: 2n = 20 (Williams, 1925).

The species is found on rock surfaces and in rock pools in the lower littoral region and also in the sublittoral. When present on the shore along with other species of *Codium* it occupies a zone below that of *Codium fragile* subsp. *tomentosoides* and above that of *C. vermilara* (as observed at Black Head, Co. Clare, in 1958).

The species is largely confined to the central and western southern coast of England and the coast of Ireland: Scilly Isles, Channel Islands, Lundy Island, Bardsey Island, Co. Cork, Co. Clare, Co. Galway, Co. Mayo, Co. Sligo, Co. Antrim, Co. Wexford, Orkney Islands, Isle of Wight, Devon, Cornwall.

There is evidence that *C. tomentosum* is dying out on the coasts of the British Isles. The record from the Orkney Islands is an old one (see Silva, 1955) and the species has not been found there in recent surveys. It was found in Dorset in 1959 (pers. record) and there are various records of its occurrence in Dorset in the early part of this century. Despite careful search it has not been seen anywhere on the Dorset coast since 1959, though two drift specimens were picked up in 1971. The nearest known population of *C. tomentosum* is at Beer just over the border in Devon. There has been a great increase in the quantity of *C. fragile* subsp. *tomentosoides* on the Dorset coast since 1959. In places it covers broad areas in the near sublittoral region. This subsp. has not been found at Beer.

Distribution records show that *C. tomentosum* has died out in Liverpool Bay. There are specimens to show that it was present in the Isle of Man at the beginning of the century and it was recorded as present by Knight & Parke (1931). The only *Codium* species there now is *C. fragile* subsp. *atlanticum*: the nearest population of *C. tomentosum* is on the Island of Bardsey. Similarly the species is disappearing from Galway Bay in Ireland. The causes of the disappearance of *C. tomentosum* are not clear. It might be related to pollution of certain sea areas or to competition with the subspecies of *C. fragile*, though the distribution of *C. fragile* subsp. *atlanticum* at its southern limit also seems to be in decline. The subject needs further investigation.

The species occurs from the Netherlands south to the Mediterranean; Azores; Sri Lanka, Japan.

Plants are present all the year round with the maximum development in winter. Gametangia can be found at almost any time of the year, but the main fruiting period seems to be over the autumn and winter months from August until the beginning of March.

The life history of *C. tomentosum* is a diploid one with the only haploid phase represented by the gametes, formed in the gametangia following meiosis. Parthenogenesis has been recorded for the species by Dangeard & Parriaud (1956), the macrogametes developing within the gametangia.

Apart from the fact that mature plants vary greatly in size, little has been recorded concerning variation for the species.

C. tomentosum does not appear to occur outside Europe and Asia. Collins (1908) commented that probably all of the records from the Pacific under the name *C. tomentosum* belong to *C. mucronatum* J. Agardh (1886).

Codium vermilara (Olivi) Chiaje (1829), p. 14.   Fig. 61E

Neotype: UC (distributed as Flora Exsiccata Austro-Hungaria No. 794, as *C. tomentosum*). Bay at Portore, Croatia.

Lamarckia vermilara Olivi (1792), p. 258, Pl. VIII.

Thallus 10–25 (–45) cm long, irregularly branched, sometimes proliferous, terete with short terminal segments; utricles obovate to oblong, square ended, with appreciable apical thickening; hairs attached towards utricle apex often forming a collar round apex rim, if attached lower, never below constriction; utricles 400–700 µm long, 100–250 µm diameter.

Gametangia oblong to fusiform (52–) 60–80 (–96) µm diameter, 200–290 µm long, 1–4 per utricle borne on protuberances ½–⅔ distance below apex.

Chromosome no.: not known.

The plants grow on rocks and in clear pools low on the shore, and in the sublittoral down to 4 m or more. They are not usually covered with epiphytes.

Apart from the Isles of Scilly and the Channel Islands, the distribution of *C. vermilara* appears to be confined to the central southern and south west counties of England and to the south, west and north coasts of Ireland. Its main occurrence is in the sublittoral region and therefore the records are probably incomplete.

The species occurs on the Atlantic shores of Europe from mid-Norway south to the Mediterranean and Adriatic Seas, but does not appear to be recorded elsewhere in the world.

The seasonal growth pattern cannot be deduced from personal observations nor from the literature. The fruiting period is also incompletely known; there are records of gametangia for August and September. The life history is not known.

Parthenogenesis has been recorded for this species by Went (1889) and Delépine (1959). The macrogametes develop within the gametangia. Gibby (1971) regards this as a consequence of the extreme conditions encountered by the species at the limits of its distribution.

CHAETOSIPHONACEAE

CHAETOSIPHONACEAE Blackman & Tansley (1902), p. 142.

Thallus of microscopic non-septate branched filaments terminating in long hairs, or of swollen multinucleate cells with one or more hairs, connected by narrow filaments; chloroplasts numerous, polygonal with single pyrenoid.

Reproduction by bi- or quadriflagellate zoospores formed in sporangium following free nuclear divisions: vegetative reproduction by formation of new coenocytes on narrow tubular outgrowths.

The family contains two genera, *Chaetosiphon* and *Blastophysa* both of which are of uncertain affinity. Only *Blastophysa* occurs in the British Isles. *Chaetosiphon* has been found only twice, by Huber (1892) and recently by O'Kelly (1980). It was at first included in the Chaetophoraceae. *Blastophysa* was also at one time classified with the Chaetophoraceae (Blackman & Tansley, 1902) and later with Valoniaceae (Wille, 1909; Collins, 1912; Printz, 1926). It is now included in the Chaetosiphonaceae along with *Chaetosiphon* because it has multinucleate cells and forms hairs. Recently O'Kelly & Yarish (1980) have suggested that *Phaeophila dendroides* might also be included in the Chaetosiphonaceae. The genus is anomalous in the Chaetophoraceae on account of the method of development of its sporangia and gametangia. In *Phaeophila dendroides* these undergo simultaneous division and become, even if only temporarily, multinucleate.

The step of transferring *Phaeophila dendroides* to the Chaetosiphonaceae has not been taken here, but the information will have to be taken into account in the future. At the present time there appears to be too little information on the other genera placed in the family.

BLASTOPHYSA Reinke

BLASTOPHYSA Reinke (1889), p. 27. (Reinke, 1888, p. 241, nomen nudum).

Type species: *B. rhizopus* Reinke (1889), p. 27

Thallus endophytic, consisting of a single multinucleate cell or series of cells, spheroidal to tubular or irregularly shaped, connected by long or short, slender or broad, colourless tubes: cells with one or a group of colourless hairs arising from upper surface; cells with a large central vacuole surrounded by a thin layer of cytoplasm with embedded nuclei and chloroplasts; chloroplast completely covering the wall or divided into polygonal plate-like segments, each with or without pyrenoid surrounded by starch sheath.

Reproduction by quadriflagellate zoospores formed in individual coenocytes following simultaneous division; vegetative reproduction by formation of new cells at ends of colourless tubes.

The genus is of uncertain affinity with only a single species.

Blastophysa rhizopus Reinke (1889), p. 27 (Reinke, 1888, p. 241, nomen nudum).

Lectotype: original illustration (Reinke, 1889, pl. 23) in the absence of material. Kieler Förde, W. Germany.

Cells irregular in shape, 20–120 μm diameter, with 1–5 colourless hairs arising from upper surface; chloroplast completely covering wall or divided into polygonal plate-like segments with or without pyrenoids; connections between cells tubular, variable in thickness, usually 10–15 μm.

Reproduction by formation of quadriflagellate zoospores in more or less rounded coenocytes: vegetative reproduction by division of coenocytes or by formation of new coenocytes from colourless tubes.

Chromosome no.: not known.

Endophytic in a variety of larger algae. Hosts recorded: *Dumontia, Nemalion, Laminaria, Ulva, Enteromorpha*; also in crust-forming algae such as *Hildenbrandtia*. Found in both the littoral and sublittoral regions.

Probably much more common in Britain than appears from the recorded distribution: Shetland Isles, Orkney Isles, Isle of Man, Norfolk, Cumbrae, Co. Antrim.
North European coasts, Mediterranean, Baltic Sea.
Southern Massachusetts (USA), Bermuda, Virgin Isles, Saba Bank.

There is very little information about the seasonal occurrence and reproduction of *Blastophysa rhizopus*. In the Isle of Man it is reported sporadically, and reproducing at all times of the year (Knight & Parke, 1931). Huber (1892) recorded zoospores in September and October in France. In the Baltic Sea it is said to be present throughout the year (Lakowitz 1929). The life history is not fully known.

The cell shape and degree of development of the colourless tubes connecting the cells appear to vary with the host species.

Parker (1970) reported cellulose I in the walls of *Blastophysa* and suggested that this indicates that it is not closely related to most of the Siphonales as delimited by Sears (1968).

GLOSSARY

ACROPETAL   The development of organs in succession towards an apex, the oldest at the base, the youngest at the tip.

ACUMINATE   Tapering gradually to a point.

ADVENTITIOUS   Arising in an irregular manner or from an abnormal position.

AKINETE   Non-motile asexual spore with a thickened wall not distinct from that of the parent cell.

ALTERNATE   Arranged in two rows, but not opposite.

ALVEOLAR   Pitted or with cavities.

ANISOGAMY   Fusion between gametes dissimilar in size, form or behaviour.

ANTHERIDIUM   Sex organ producing flagellated male gametes.

ANTHEROZOID   Flagellate male gamete.

APICULATE   Narrowing suddenly to a broad point at the apex.

APLANOSPORE   Non-motile asexual spore having a wall distinct from that of the parent cell.

APPRESSED   Pressed close.

ARCHAEOPLASTID   Chloroplast in the form of a perforated plate.

ARCUATE   Curved like the arc of a circle.

ATTENUATE   Tapering gradually.

AXIS   The central or median plane of a cell or plant.

BASIONYM   Name-bringing or epithet-bringing synonym.

BASIPETAL   The development of organs in succession towards the base, the youngest at the base, the oldest at the tip.

BILENTICULAR PYRENOID   Pyrenoid surrounded by only two plates of starch.

BULLATE   With the surface appearing swollen and blistered.

CAESPITOSE   Tufted.

CHLOROPLAST   Green chlorophyll-containing plastid.

CLAVATE   Club-shaped.

COENOBIUM   Colony of algal cells of a definite cell arrangement and number not increasing at maturity.

COENOCYTE   A multinucleate cell or thallus in which there are no dividing walls.

COMPLANATE   More or less in one plane.

CORTEX   A term used loosely for the peripheral region of a thallus, but not including the epidermis if one is present.

CORYMBOSE   A flat-topped branch system.

CRENATE   With a notched edge, with obtuse or flattened teeth.

CRUSTOSE   Forming a crust, growing flattened against a substrate.

CUNEATE   Wedge-shaped.

DECUMBENT   Growing horizontally but with extremities ascending.

DENDROID   Tree-like.

DENTATE   With tooth-like projections.

DIOECIOUS   With male and female gametangia on separate plants.

DISCOID   Flat and circular.

DISTAL   At the end remote from the base or point of attachment.

DISTICHOUS   Arranged in two opposite rows.

DISTROMATIC Formed of two layers of cells.

DIVARICATE Spreading widely.

ENDOPHYTIC Living within a plant.

ENDOZOIC Living within an animal.

ENTIRE Without lobes or divisions.

EPILITHIC Living on rock or stones.

EPIPHYTIC Living on a plant but attached to the surface only.

EPIZOIC Living on an animal, but attached to the surface only.

EVECTION The development of lateral branches to form pseudodichotomies by differential growth of the wall of the parent axis.

FALCATE Sickle-shaped.

FASCICLE A bundle or cluster; adjective fasciculate.

FASTIGIATE Of a branching system, conical or pyramidal in outline, with either the base or apex uppermost.

FENESTRATE Irregularly perforate.

FILAMENT A branched or unbranched row of cells joined end to end.

FILIFORM Thread-like.

FLABELLATE Fan-shaped.

FLACCID Limp.

FLAGELLUM A thread-like cytoplasmic outgrowth used for locomotion.

FOLIOSE With a broad, flat blade.

FROND That part of a thallus other than the attachment structure.

FURCATE Forked.

FUSIFORM Spindle-shaped.

GAMETANGIUM A cell producing one or more gametes.

GAMETE A sexual cell capable of uniting with another sexual cell.

GAMETOPHYTE Thallus, either haploid or diploid which bears gametangia.

HAIR A colourless, nucleate, vegetative cell with a narrow extension.

HAPTERON A specialised multicellular attachment structure; pl. haptera.

HETEROPLASTIC Having both chloroplasts and colourless amyloplasts.

HOLOTYPE The one specimen or other element used by an author or designated by him as the nomenclatural type.

HOMOPLASTIC Having only one type of plastid, the chloroplast.

LANCEOLATE Narrow and tapering at both ends.

LECTOTYPE A specimen or other element selected from the original material to serve as a nomenclatural type when no holotype was designated at the time of publication or for as long as it is missing.

LIGULATE Strap-shaped.

LINEAR Long and narrow with parallel sides.

LITTORAL Applied to that part of the shore which is alternately exposed to the air and wetted, either by the tide or by splash or spray (see p. 5).

LUBRICOUS Slippery.

MEDULLA Central region, filamentous, pseudoparenchymatous, or parenchymatous in construction, of an internally differentiated thallus.

MEMBRANOUS Forming a thin layer.

MESOPLASTID Chloroplast made up of anastomosing, but originally separate, plates.

MONOECIOUS With male and female gametangia on the same thallus.

MONOSTROMATIC One cell in thickness, with reference to a lamina or disc.

MUCRONATE Narrowing suddenly to a short narrow point, cf. acuminate and apiculate.

OOGAMY Fusion between a male gamete and a non-motile female gamete.

ORBICULATE Circular in outline.

PAPILLA A short nipple-like outgrowth.

PARENCHYMA A compact tissue formed by cell division in all planes.

PARIETAL Lying along the wall.

PATENT Spreading.

PECTINATE With narrow branches set along an axis like the teeth of a comb.

PEDICEL A stalk of a reproductive organ.

PEDICELLATE Stalked.

PELTATE More or less circular and attached by the centre of the lower surface.

PILIFEROUS Bearing hairs.

PLASMODESMATA Strands passing through cell walls connecting protoplasts of adjacent living vegetative cells.

PLASTID Cell inclusion, either pigmented (chromatophore) or unpigmented (leucoplast).

POLYPYRAMIDAL PYRENOID Pyrenoid surrounded by many plates of starch.

POLYSTROMATIC Formed of several or many layers of cells.

PROLIFERATION The formation of new branches by vegetative cell division.

PSEUDOPARENCHYMA A tissue formed by the aggregation of branched or unbranched filaments and having the appearance of parenchyma.

PULVINATE Cushion-shaped.

PUNCTATE Marked with points, dots, spots or depressions resembling punctures.

PYRENOID An organelle occurring within or adjacent to a chloroplast; often associated with reserve food accumulation.

PYRIFORM Pear-shaped.

QUADRATE Square.

RAMULUS A small branch.

REFLEXED Bent or turned back.

REMOTE Situated at a distance or interval from one another.

RENIFORM Kidney-shaped.

RETICULATE Having the appearance of a network.

RHIZOID A unicellular or multicellular filament produced either externally (for attachment) or internally.

SACCULATE Regularly inflated.

SECUND Arranged on one side only.

SERIATE Regularly arranged in one or more series.

SERRATE Having small projections like the teeth of a saw.

SESSILE Without a stalk.

SETA A hyaline hair-like outgrowth of a vegetative cell which lacks a nucleus.

SETACEOUS Bristle-like.

SINUATE Of a plane structure with a wavy margin.

SIPHONACEOUS Tubular, non-cellular filament. Characteristic structure found in the Codiales.

SORUS An aggregation of reproductive structures.

SPATHULATE Spoon-shaped.

SPORANGIUM An organ producing one or more spores.

STIPE The lower-most, stalk-like portion of an erect frond.

SUBLITTORAL Applied to that part of the shore which is either totally immersed or only uncovered by the receding tide infrequently and then only for very short periods (see p. 5).

SWARMER A motile reproductive body of uncertain function.

TAXON A taxonomic group of any rank.

TERETE Circular in transverse section.

THALLUS A plant body of relatively simple organisation.

TOMENTOSE Closely covered with short hairs.

TRABECULAE Filaments traversing a cavity.

TUBULAR Cylindrical and hollow.

UNDULATE Having a wavy surface.

UNISERIATE With cells arranged in a single row.

UTRICLE Swollen bladder-like structure formed from a single cell or coenocyte.

ZOOSPORE Motile asexual reproductive cell.

ZYGOTE Cell formed after fusion of two gametes.

REFERENCES

ABBOTT, I. A. & HOLLENBERG, G. J. 1976. *Marine Algae of California.* Stanford.

ADANSON, M. 1763. *Familles des plantes* II Partie. Paris.

AGARDH, C. A. 1817. *Synopsis Algarum Scandinaviae.* Lundae.

AGARDH, C. A. 1822. *Species Algarum* 1(2). Lundae.

AGARDH, C. A. 1824. *Systema Algarum.* Lundae.

AGARDH, J. G. 1842. *Algae Maris Mediterranei et Adriatici.* Parisiis.

AGARDH, J. G. 1848. *Anadema*, ett nytt slägte bland Algerne; beskrivet af J. G. Agardh. *Kongl. Vetensk. Akad. Handl.* 1846: 1–16.

AGARDH, J. G. 1882–83. Till Algernes Systematik. . . . (Tredja afdelningen). *Acta Univ. Lund.* 19(2): 1–182.

AGARDH, J. G. 1887. Till Algernes Systematik. . . . (Femte afdelningen). VIII. Siphoniae. *Acta Univ. Lund.* 23(2): 1–180.

ALEEM, A. A. & SCHULZ, E. 1952. Über Zonierung von Algengemeinschaften. (Ökologische Untersuchungen im Nord-Ostsee-Kanal, I.) *Kieler Meeresforsch.* 9: 70–6.

ÅLVIK, G. 1934. Plankton-Algen norwegischer Austernpollen. *Bergens Mus. Årb.* 1934(6): 1–47.

ANAND, P. L. 1937. A taxonomic study of the algae of the British chalk-cliffs. *J. Bot., Lond.* 75(Suppl. II): 1–51.

ARASAKI, S. & SHIHIRA, I. 1959. Variability of morphological structure and mode of reproduction in *Enteromorpha linza. Jap. J. Bot.* 17: 92–100.

ARCHER, A. A. 1963. *A New Approach to the Taxonomy of the Branched Members of the Cladophoraceae in the British Isles.* Thesis Liverpool University.

ARESCHOUG, J. E. 1850. *Letterstedtia*, ny alg-form från Port Natal. *Ofvers. K. Vetensk. Akad. Forh. Stockh.* 7: 1–4.

ARESCHOUG, J. E. 1850a. Phycearum, quae in maribus Scandinaviae crescunt, enumeratio. Sectio posterior Ulvaceas continens. *Nova Acta R. Soc. Scient. Upsal.* 14: 385–454.

ARESCHOUG, J. E. 1866. Observationes Phycologicae. Particula prima. De Confervaceis nonnullis. *Nova Acta R. Soc. Sci. Upsal.* Ser. 3, 6: 1–26.

ARESCHOUG, J. E. 1874. Observationes Phycologicae. Particula secunda. De *Urospora mirabili* Aresch. et de Chloro-zoosporarum copulatione. *Nova Acta R. Soc. Sci. Upsal.* Ser. 3, 9: 1–13.

BARTLETT, R. B. & SOUTH, G. R. 1973. Observations on the life-history of *Bryopsis hypnoides* Lamour. from Newfoundland: a new variation in culture. *Acta Bot. Neerl.* 22: 1–5.

BATTERS, E. A. L. 1895. Some new British algae. *Ann. Bot.* 9: 168–169.

BATTERS, E. A. L. 1900. New or critical British marine algae. *J. Bot., Lond.* 38: 369–379.

BATTERS, E. A. L. 1902. A catalogue of the British marine algae. *J. Bot., Lond.* 40 (Suppl.): 1–107.

BERGER-PERROT, Y. 1980a. *Ulothrix flacca* (Dillwyn) Thuret (Chlorophycée, Ulotrichale) des côtes de Bretagne et son polymorphisme. *Cryptogamie Algol.* 1: 229–248.

BERGER-PERROT, Y. 1980b. Recherches sur l'*Ulothrix flacca* (Dillwyn) Thuret (Chlorophycée, Ulotrichale) des côtes de Bretagne. I – Morphologie, cytologie, caryologie et réproduction de la variété *geniculata* (Jónsson) Berger-Perrot. *Cryptogamie Algol.* 1: 327–354.

BERGER-PERROT, Y. 1980c. Trois nouvelles espèces d'*Urospora* à cellules uninuclées sur les côtes de Bretagne. *Cryptogamie Algol.* 1: 141–160.

BERGER–PERROT, Y. 1981. Mise sur le problème concernant l'*Ulothrix speciosa* (Carm. ex Harvey) Kützing et l'*Urospora kornmanii* Berger-Perrot. Etude comparée de la réproduction et du cycle de développement des deux espèces sur les côtes de Bretagne. *Phycologia* 20: 147–164.

BERNATOWICZ, A. J. 1958. Ecological isolation and independent speciation of the alternate generations of plants. *Biol. Bull. Mar. Biol. Lab. Woods Hole* 115: 323.

BLACKMANN, F. F. & TANSLEY, A. G. 1902. A revision of the classification of the green algae. *New Phytol.* **1**: 89–96, 133–144.

BLIDING, C. 1936. Über die Fortpflanzungskörper einiger mariner *Cladophora*-Arten. *Svensk Bot. Tidskr.* **30**: 529–536.

BLIDING, C. 1944. Zur Systematik der schwedischen Enteromorphen. *Bot. Notiser* **1944**: 331–356.

BLIDING, C. 1955. *Enteromorpha intermedia.* A new species from the coasts of Sweden, England and Wales. *Bot. Notiser* **108**: 253–262.

BLIDING, C. 1957. Studies in *Rhizoclonium* I. Life history of two species. *Bot. Notiser* **110**: 271–275.

BLIDING, C. 1960. A preliminary report on some new Mediterranean green algae. *Bot. Notiser* **113**: 172–184.

BLIDING, C. 1963. A critical survey of European taxa in Ulvales. Part I *Capsosiphon, Percursaria, Blidingia, Enteromorpha. Op. Bot. Soc. Bot. Lund* **8**: 1–160.

BLIDING, C. 1968. A critical survey of European taxa in Ulvales, II *Ulva, Ulvaria, Monostroma, Kornmannia. Bot. Notiser* **121**: 535–629.

BÖHLKE, K. 1978. The biology of some members of the genus *Chlorochytrium* Cohn. Thesis, Liverpool University.

BOLD, H. C. & WYNNE, M. J. 1978. *Introduction to the Algae. Structure and Reproduction.* New Jersey.

BØRGESEN, F. 1902. Marine Algae, pp. 339–532. *In*, Warming, E. (Ed.), *Botany of the Faröes Based Upon Danish Investigations.* Part II. Copenhagen.

BORNET, E. & FLAHAULT, C. 1888. Note sur deux nouveaux genres d'algues perforantes. *J. Bot., Paris* **2**: 161–165.

BORNET, E. & FLAHAULT, C. 1889. Sur quelques plantes vivant dans le test calcaire des mollusques. *Bull. Soc. Bot. Fr.* **36**: CXLVII–CLXXVI.

BORNET, E. & THURET, G. 1880. *Notes algologiques.* Fascicule II. Paris.

BORY DE ST VINCENT, J. B. G. M. 1823. Confervées. *Dictionnaire Classique d'Histoire Naturelle.* **4**. Paris.

BORY DE ST VINCENT, J. B. G. M. 1829. Cryptogamie, pp. 201–250. *In*, Duperrey, L. I. (Ed.), *Voyage autour du monde … La Coquille … 1822–25.* Paris.

BORZI, A. 1883. *Studi Algologici Saggio di Ricerche Sulla Biologia delle Alghe.* Fascicolo I. Messina.

BORZI, A. 1895. *Studi Algologici Saggio di Ricerche Sulla Biologia delle Alghe.* Fascicolo II. Palermo.

BRAUN, A. 1855. *Algarum Unicellularum Genera Nova et Minus Cognita.* Lipsiae.

BRAVO, L. M. 1962. A contribution to knowledge of the life history of *Prasiola meridionalis. Phycologia* **2**: 17–23.

BRAVO, L. M. 1965. Studies on the life history of *Prasiola meridionalis. Phycologia* **4**: 177–194.

BRISTOL, B. M. 1920. A review of the genus *Chlorochytrium* Cohn. *Bot. J. Linn. Soc.* **45**: 1–26.

BURROWS, E. M. 1959. Growth form and environment in *Enteromorpha. Bot. J. Linn. Soc.* **56**: 204–206.

BURROWS, E. M. 1964. A preliminary list of the marine algae of the coast of Dorset. *Br. Phycol. Bull.* **2**: 364–368.

CABRERA, A. 1823. Descriptio novi generis Algarum. *Physiogr. Sällsk. Årsb.* **1823**: 99.

CARTER, N. 1926. An investigation into the cytology and biology of the Ulvaceae. *Ann. Bot.* **40**: 665–689.

CARTER, N. 1933. A comparative study of the alga flora of two saltmarshes. Part II. *J. Ecol.* **21**: 128–208.

CHADEFAUD, M. 1957. Sur l'*Enteromorpha chadefaudii* J. Feldmann. *Rev. gen. Bot.* **64**: 653–669.

CHAPMAN, V. J. 1949. Some new species and forms of marine algae from New Zealand. *Farlowia* **3**: 495–498.

CHAPMAN, V. J. 1952. New entities in the Chlorophyceae of New Zealand. *Trans. R. Soc. N.Z.* **80**: 47–58.

CHAPMAN, V. J. 1956. The marine algae of New Zealand. Part I. Myxophyceae and Chlorophyceae. *Bot. J. Linn. Soc.* **55**: 333–501.

CHIAJE, S. DELLE 1829. *Hydrophytologiae Regni Neapolitani.* Neapoli.

CHIHARA, M. 1962. Occurrence of the *Gomontia*-like phase in the life history of certain species belonging to *Collinsiella* and *Monostroma* (a preliminary note). *J. Jap. Bot.* **37**: 44–45.

CHIHARA, M. 1967. Developmental morphology and systematics of *Capsosiphon fulvescens* as found in Izu, Japan. *Bull. Natn. Sci. Mus. Tokyo* **10**: 163–170.

CHODAT, R. 1897. Sur deux algues perforantes de l'ile de Man. *Bull. Herb. Boissier* **5**: 712–716.

CHRISTENSEN, T. 1957a. Some Irish algal finds. *Br. Phycol. Bull.* **1**: 42–43.

CHRISTENSEN, T. 1957b. *Chaetomorpha linum* in the attached state. *Bot. Tidsskr.* **53**: 311–316.

CHRISTENSEN, T. 1985. *Microspora ficulinae*, a green alga living in marine sponges. *Br. Phycol. J.* **20**: 5–7.

CHRISTENSEN, T. & THOMSEN, H. A. 1974. *Algefortegnelse. Oversigt Over Udbredelsen af Danske Salt- og Brakvandsarter Fraset ikke- planktiske Kiselalger.* København.

COHN, F. 1872. Ueber parasitische Algen. *Beitr. Biol. Pfl.* **1**(2): 87–108.

COLE, K. & AKINTOBI, S. 1963. The life cycle of *Prasiola meridionalis* Setchell and Gardner. *Can. J. Bot.* **4**: 661–668.

COLLINS, F. S. 1908. Notes on algae. IX. *Rhodora* **10**: 155–164.

COLLINS, F. S. 1909. The green algae of North America. *Tufts Coll. Stud.* **2**: 79–490.

COLLINS, F. S. 1912. The green algae of North America, supplementary paper. *Tufts Coll. Stud.* **3**: 69–109.

COMPS, B. 1960. A propos des phénomènes nucléaires de la reproduction chez *Enteromorpha linza* (L.) J. Ag. *C. r. hebd. Seanc. Acad. Sci., Paris* **251**: 2067–2069.

COTTON, A. D. 1912. Marine Algae. *In*, Praeger R. L. (Ed.), A biological survey of Clare Island in the county of Mayo, Ireland and of the adjoining district. *Proc. R. Ir. Acad.* **31** sect. 1(15): 1–178.

COX, E. J. 1975. Further studies on the genus *Berkeleya* Grev. *Br. Phycol. J.* **10**: 205–217.

CROUAN, P. L. & CROUAN, H. M. 1852. *Algues Marines du Finistère. 3 Zoospermées.* Brest.

CROUAN, P. L. & CROUAN, H. M. 1859. Notice sur quelques espèces et genres nouveaux d'algues marines de la Rade de Brest. *Annls Sci. Nat. Bot.* Sér. 4, **12**: 288–292.

CROUAN, P. L. & CROUAN, H. M. 1867. *Florule du Finistère.* Paris & Brest.

DANGEARD, P. A. 1912. Recherches sur quelques algues nouvelles ou peu connues. *Le Botaniste* **12**: I–XXVI.

DANGEARD, P. 1931. L'*Ulvella lens* de Crouan et l'*Ulvella setchellii* sp. nov. *Bull. Soc. Bot. Fr.* **78**: 312–318.

DANGEARD, P. 1951. Sur une espèce d'*Ulva* de nos côtes atlantiques (*U. olivacea* n. sp.). *Le Botaniste* **35**: 27–34.

DANGEARD, P. 1958. La reproduction et le développement de l'*Enteromorpha marginata* Ag. et le rattachement de cette espèce au genre *Blidingia*. *C. r. hebd. Séanc. Acad. Sci., Paris* **246**: 347–351.

DANGEARD, P. 1961. Le problème de l'espèce avec référence au groupe des Ulvacées. *Le Botaniste* **44**: 21–36.

DANGEARD, P. & PARRIAUD, H. 1956. Sur quelques cas de développement apogamique chez deux espèces de *Codium* de la région du sudouest. *C. r. hebd. Séanc. Acad. Sci., Paris* **243**: 1981–1983.

DAVIS, P. H. & HEYWOOD, V. H. 1963. *Principles of Angiosperm Taxonomy.* Edinburgh & London.

DELÉPINE, R. 1959. Observations sur quelques *Codium* (Chlorophycées) des côtes Françaises. *Revue Gén. Bot.* **66**: 366–394.

DELLOW, V. 1952. The genus *Codium* in New Zealand. Part I. Systematics. *Trans. R. Soc. N.Z.* **80**: 119–141.

DE NOTARIS, G. 1846. *Prospetto della Flora Ligustica e dei Zoofiti del Mare Ligustico.* Genova.

DE TONI, J. B. 1889. *Sylloge Algarum.* 1. Patavii.

DE TONI, G. B. & LEVI, D. 1888. Flora algologia della Venezia. *Atti R. Inst. Veneto Sci.* Ser. 6, **6**: 95–155; 289–350.

DE VALERA, M. 1939. Some new or critical algae from Galway Bay, Ireland. *K. fysiogr. Sällsk. Lund Förh.* **9**: 91–104.

DIAZ-PIFERRER, M. 1972. *A Taxonomic Study of the Genus* Bryopsis *in the British Isles and in the Caribbean.* Thesis, Liverpool University.

DICKIE, G. 1870. Notes on some algae found in the North-Atlantic Ocean. *Bot. J. Linn. Soc.* **11**: 456–459.

DICKINSON, C. I. 1963. *British Seaweeds.* Frome & London.

DILLENIUS, J. J. 1742. *Historia Muscorum.* Oxford.

DILLWYN, L. W. 1802–09. *British Confervae.* London.

DOTY, M. S. 1947. The marine algae of Oregon. Part I. Chlorophyta and Phaeophyta. *Farlowia* **3**: 1–65.

DUBE, M. A. 1967. On the life history of *Monostroma fuscum* (Postels et Ruprecht) Wittrock. *J. Phycol.* **3**: 64–73.

DUCKER, S. C. 1958. A new species of *Basicladia* on Australian freshwater turtles. *Hydrobiologia* **10**: 157–174.

DUMORTIER, B. C. 1823. *Commentationes Botanicae. Observations botaniques.* Tournay.

EATON, J. W., BROWN, J. G. & ROUND, F. E. 1966. Some observations on polarity and regeneration in *Enteromorpha. Br. Phycol. Bull.* **3**: 53–62.

EDELSTEIN, T. & MCLACHLAN, J. 1966. Investigations of the marine algae of Nova Scotia. I. Winter flora of the Atlantic coast. *Can. J. Bot.* **44**: 1035–1055.

EDWARDS, P. 1975a. Evidence for a relationship between the genera *Rosenvingiella* and *Prasiola* (Chlorophyta). *Br. Phycol. J.* **10**: 291–297.

EDWARDS, P. 1975b. An assessment of possible pollution effects over a century on the benthic marine algae of Co. Durham, England. *Bot. J. Linn. Soc.* **70**: 269–305.

ENDLICHER, S. L. 1843. *Mantissa Altera Sistens Generum Plantarum.* Suppl. III. Vindobonae.

FALKENBERG, P. 1879. Die Meeres–Algen des Golfes von Neapel. *Mitt. Zool. Stn Neapel* **1**: 218–277.

FAN, K. C. 1957. Observations on the life-history of *Codiolum petrocelidis. Phycol. Newsl.* **10**(3): 75.

FAN, K. C. 1959. Studies on the life histories of marine algae I. *Codiolum petrocelidis* and *Spongomorpha coalita. Bull. Torrey Bot. Club* **86**: 1–12.

FARLOW, W. G. 1881. *Marine Algae of New England and Adjacent Coast.* Washington.

FARNHAM, W. F., BLUNDEN, G. & GORDON, S. M. 1985. Occurrence and pigment analysis of the sponge endobiont *Microspora ficulinae* (Chlorophyceae). *Botanica mar.* **28**: 79–81.

FAROOQUI, P. B. 1969. A note on the genus *Chlorochytrium* Fott (Ulotrichaceae). *Preslia* **41**: 1–7.

FELDMANN, J. 1937. Les algues marines de la côte des Albères. I–III Cyanophycées, Chlorophycées, Phaeophycées. *Revue algol.* **9**: 141–335.

FELDMANN, J. 1950. Sur l'existence d'une alternance de générations entre l'*Halicystis parvula* Schmitz et le *Derbesia tenuissima* (De Not.) Crn. *C. r. hebd. Séanc. Acad. Sci., Paris* **230**: 322–323.

FELDMANN, J. 1952. Les cycles de reproduction des algues et leurs rapports avec le phylogénie. *Revue Cytol. Biol. Vég.* **13**: 1–49.

FELDMANN, J. 1954a. Inventaire de la flore marine de Roscoff. *Trav. Stat. Biol. Roscoff* N.S.5, Suppl. 6: 1–152.

FELDMANN, J. 1954b. *Huitième Congrès International de Botanique, Paris 1954. Rapports et communications parvenus avant le congrès la section 17.* Paris.

FELDMANN, J. 1956. Sur la parthénogénèse du *Codium fragile* (Sur.) Hariot dans la Méditerranée. *C. r. hebd. Séanc. Acad. Sci., Paris* **243**: 305–307.

FLOYD, G. L., STEWART, K. D. & MATTOX, K. R. 1971. Cytokinesis and plasmodesmata in *Ulothrix. J. Phycol.* **7**: 306–309.

FLOYD, G. L., STEWART, K. D. & MATTOX, K. R. 1972a. Comparative cytology of *Ulothrix* and *Stigeoclonium. J. Phycol.* **8**: 68–81.

FLOYD, G. L., STEWART, K. D. & MATTOX, K. R. 1972b. Cellular organisation, mitosis and cytokinesis in the Ulotrichalean alga, *Klebsormidium*. *J. Phycol.* **8**: 176–184.

FOSLIE, M. 1887. Nye havsalger. *Tromsø Mus. Aarsh.* **10**: 175–195.

FOSLIE, M. 1890. Contributions to knowledge of the marine algae of Norway. I. East-Finmarken. *Tromsø Mus. Aarsh.* **13**: 1–186.

FÖYN, B. 1929. Untersuchungen über die sexualität und Entwicklung von algen. IV. Vorläufige Mitteilung über die Sexualität und den Generationswechsel von *Cladophora* und *Ulva*. *Ber. dt. Bot. Ges.* **47**: 495–506.

FÖYN, B. 1934. Lebenszyklus, Cytologie und Sexualität der Chlorophycee *Cladophora suhriana* Kützing. *Arch. Protistenk.* **83**: 1–56.

FRIEDMANN, I. 1959. Structure, life-history, and sex determination of *Prasiola stipitata* Suhr. *Ann. Bot.* N.S. **23**: 571–594.

FRIEDMANN, I. 1960. Gametes, fertilization and zygote development in *Prasiola stipitata* Suhr. I. Light microscopy. *Nova Hedwigia* **1**: 333–344.

FRIEDMANN, I. 1969. Geographic and environmental factors controlling life history and morphology in *Prasiola stipitata* Suhr. *Öst. Bot. Z.* **116**: 203–225.

FRIES, E. 1825. *Systema orbis vegetabilis. Primas lineas novae constructionis perclitatur E. Fries. Pars. i. Plantae Homonemeae*. Lund.

FRIES, E. 1835. *Corpus Florarum Provincialum Sueciae. I. Floram Scanicam*. Upsaliae.

FRITSCH, F. E. 1935. *The Structure and Reproduction of the Algae*. Vol. **1**. London & New York.

FUNK, G. 1927. Die Algenvegetation des Golfs von Neapel. *Pubbl. Staz. Zool. Napoli* **7** (Suppl.): 1–507.

FUNK, G. 1955. Beiträge zur kenntnis der Meeresalgen von Neapel zugleich mikrophotographischer Atlas. *Pubbl. Staz. Zool. Napoli* **25** (Suppl.): 1–178.

GAILLON, B. 1828. Thalassiophytes. *In, Dictionnaire des sciences naturelles.* **53**. Strasbourg & Paris.

GAILLON, B. 1833. *Aperçu d'Histoire Naturelle et Observations sur les Limites qui Séparent le Règne Végétal du Règne Animal*. Boulogne.

GALLAGHER, S. B. & HUMM, H. J. 1980. *Pilinia earleae* (Chlorophyceae, Chroolepidaceae) from the Florida west coast. *J. Phycol* **16**: 532–536.

GARDNER, N. L. 1917. New Pacific coast marine algae I. *Univ. Calif. Publs Bot.* **6**: 377–416.

GAYRAL, P. 1962. Sur l'existence de gamétophytes chez '*Ulva olivascens*' P. Dang. du détroit de Gibraltar, description du développement de cette espèce. *Le Botaniste* **45**: 83–91.

GAYRAL, P. 1964. Sur le démembrement de l'actuel genre *Monostroma* Thuret (Chlorophycées, Ulotrichales s.l.). *C. r. hebd. Séanc. Acad. Sci., Paris* **258**: 2149–2152.

GAYRAL, P. 1971. Mise au point sur la systématique de l'ordre des Ulvales. *Bull. Soc. Phycol. Fr.* **16**: 63–67.

GAYRAL, P. & LEPAILLEUR, H. 1968. Une rare Chlorophycée unicellulaire marine: *Heterogonium salinum* P. A. Dangeard. *Rev. algol.* N.S. **9**: 131–134.

GEORGE, E. A. 1957. A note on *Stichococcus bacillaris* Naeg. and some species of *Chlorella* as marine algae. *J. Mar. Biol. Ass. U.K.* **36**: 111–114.

GIBBY, J. M. 1971. *The Present Status of* Codium *in Britain*. Thesis, Liverpool University.

GOBI, C. 1879. Berichte über die algologischen Forschungen in finnischen Meerbusen im Sommer 1877 ausgeführt. *Trudy Leningr. Obshch. Estest.* **10**: 83–92.

GODWARD, M. B. E. 1966. *The Chromosomes of the Algae*. London.

GOSS-CUSTARD, S., JONES, J., KITCHING, J. A. & NORTON, T. A. 1979. Tide pools of Carrigathorna and Barloge Creek. *Phil. Trans. R. Soc.* B, **287**: 1–44.

GREVILLE, R. K. 1824. *Flora Edinensis*. Edinburgh.

GREVILLE, R. K. 1826. *Scottish Cryptogamic Flora*. **4**. Edinburgh.

GREVILLE, R. K. 1830. *Algae Britannicae*. Edinburgh.

HAMEL, A. & HAMEL, G. 1929. Sur l'hétérogamie d'une Cladophoracée *Lola* (nov. gen.) *librica* (Setch. et Gardn.). *C. r. hebd. Séanc. Acad. Sci., Paris* **189**: 1094–1096.

HAMEL, G. 1924. Quelques *Cladophora* des côtes françaises. *Revue algol.* **1**: 168–174.

HAMEL, G. 1930. Chlorophycées des côtes françaises. *Revue Algol.* **5**: 1–54.

HAMEL, G. 1931a. Chlorophycées des côtes françaises. *Revue Algol.* **5**: 383–430.

HAMEL, G. 1931b. Chlorophycées des côtes françaises. *Revue Algol.* **6**: 9–73.

HANIC, L. 1965. *Life-history Studies on* Urospora *and* Codiolum *from Southern British Columbia.* Thesis, University of British Columbia.

HANIC, L. A. 1979. Observations on *Prasiola meridionalis* (S. & G.) and *Rosenvingiella constricta* (S. & G.) Silva (Chlorophyta, Prasiolales) from Galiano Island, British Columbia. *Phycologia* **18**: 71–76.

HANSGIRG, A. 1886. *Physiologische und algologische Studien.* Prag.

HARIOT, P. 1889. Algues. *In, Mission scientifique du Cap Horn, 1882–1883* **5**. Botanique. Paris.

HARTMANN, M. 1929. Untersuchungen über die Sexualität und Entwicklung von Algen. III. Über die sexualität und Generationswechsel von *Chaetomorpha* und *Enteromorpha. Ber. dt. Bot. Ges.* **47**: 485–494.

HARTOG, C. DEN 1959. The epilithic algal communites occurring along the coasts of the Netherlands. *Wentia* **1**: 1–241.

HARVEY, W. H. 1833. Div. II. Confervoideae. Div. III. Gloiocladeae. *In*, Hooker, W. J. (Ed.) *The English flora of Sir James Edward Smith* **5**. London.

HARVEY, W. H. 1846–51. *Phycologia Britannica.* London. [Includes references dated 1846, 1848, 1849, 1851.]

HARVEY, W. H. 1849a. *A Manual of the British Marine Algae.* London.

HAUCK, F. 1876. Verzeichniss der im Golfe von Triest gesammelten Meeralgen. (Fortsetzung). *Öst. Bot. Z.* **26**: 54–57.

HAYWOOD, J. 1974. Studies on the growth of *Stichococcus bacillaris* Naeg. in culture. *J. Mar. Biol. Ass. U.K.* **54**: 261–268.

HAZEN, T. E. 1902. The Ulothricaceae and Chaetophoraceae of the United States. *Mem. Torrey Bot. Club* **11**: 135–250.

HEYNHOLD, G. 1841. *Nomenclator Botanicus Hortensis* **1**. Dresden & Leipzig.

HIROSE, H. & YOSHIDA, K. 1964. A review of the life history of the genus *Monostroma. Bull. Jap. Soc. Phycol.* **12**: 19–31.

HOFFMAN, W. E. & TILDEN, J. E. 1930. *Basicladia*, a new genus of Cladophoraceae. *Bot. Gaz.* **89**: 374–384.

HÖHNEL, F. 1920. Mykologische Fragmente. *Annls. Mycol.* **18**: 71–97.

HOLLENBERG, G. J. 1957. Culture studies of *Spongomorpha coalita. Phycol. Newsl.* **32**: 76.

HOLLENBERG, G. J. 1958. Observations concerning the life cyle of *Spongomorpha coalita* (Ruprecht) Collins. *Madroño* **14**: 249–251.

HOLMES, E. M. & BATTERS, E. A. L. 1890. A revised list of British marine algae. *Ann. Bot., Lond.* **5**: 63–107.

HOOKER, W. J. 1833. *British Flora* **2**. London.

HOOPER, R. & SOUTH, G. R. 1977. Additions to the benthic marine algal flora of Newfoundland III, with observations on species new to Eastern Canada and North America. *Naturaliste Can.* **104**: 383–394.

HORI, T. 1972. Ultrastructure of the pyrenoid of *Monostroma* (Chlorophyceae) and related genera, pp. 17–32. *In*, Abbott, I. A. & Kurogi, M. (Eds), *Contributions to the Systematics of Benthic Marine Algae of the North Pacific.* Kobe.

HORNEMANN, I. M. 1813. *Flora Danica* **9**, Fasc. 25. Hauniae.

HORNEMANN, I. M. 1816. *Flora Danica* **9**, Fasc. 26. Hauniae.

HUBER, J. 1892. Contributions à la connaissance des Chaetophorées epiphytes et endophytes et de leurs affinités. *Ann. Sci. Nat., Bot.* Ser. 7, **16**: 265–359.

HUDSON, W. 1762. *Flora Anglica.* London.

HUDSON, W. 1778. *Flora Anglica* ed. 2. London.

HUMM, H. J. & TAYLOR, S. E. 1961. Marine Chlorophyta of the upper west coast of Florida. *Bull. Mar. Sci. Gulf Caribb.* **11**: 321–380.

HYGEN, G. 1937. Über Gametenkopulation bei der Grünalge *Bolbocoleon piliferum* Pringsh. *Nytt Mag. Naturvid.* **77** 133–135.

IRVINE, D. E. G. & JOHN, D. M. 1985. *Systematics of the Green Algae*. Systematics Association Special Volume **27**. London.

IRVINE, D. E. G., GUIRY, M. D., TITTLEY, I. & RUSSELL, G. 1975. New and interesting marine algae from the Shetland Isles. *Br. Phycol. J.* **10**: 57–71.

IWAMOTO, K. 1959. On the green algae growing on the Nori-culture-net in Tokyo Bay. *Bull. Jap. Soc. Phycol.* **7**: 4–11.

JAASUND, E. 1965. Aspects of the vegetation of north Norway. *Botanica Gothoburg.* **4**: 1–174.

JESSEN, C. F. G. 1848. *Prasiolae Generis Algarum Monographia. Dissertatio Inauguralis Botanica.* Kiliae.

JÓNSSON, H. 1912. The marine algal vegetation of Iceland. *In*, Rosenvinge, I. K. & Warming E. (Eds), *The Botany of Iceland*. Copenhagen & London.

JÓNSSON, S. 1959. Le cycle de développement du *Spongomorpha lanosa* (Roth) Kütz. et la nouvelle famille des Acrosiphoniacées. *C. r. hebd. Séanc. Acad. Sci., Paris* **248**: 1565–1567.

JÓNSSON, S. 1962. Recherches sur des Cladophoracées marines structure, reproduction, cycles comparés, consequences systématiques. *Ann. Sci. Nat. Bot.* Sér. 12, **5**: 25–264.

JÓNSSON, S. 1964. Existence d'une caryogamie facultative chez l'*Acrosiphonia spinescens* (Kütz.) Kjellm. *C. r. hebd. Séanc. Acad. Sci., Paris* **258**: 6207–6209.

JÓNSSON, S. & PERROT, Y. 1967. Le cycle de reproduction du *Cladophora rupestris* (L.) Kütz. (Cladophoracees). *C. r. hebd. Séanc. Acad. Sci., Paris* Sér. D, **264**: 2628–2631.

JORDE, I. 1933. Untersuchungen über den Lebenszyklus von *Urospora* Aresch. und *Codiolum* A. Braun. *Nyt Mag. Naturvid.* **73**: 1–20.

KAPRAUN, D. F. 1970. Field and cultural studies of *Ulva* and *Enteromorpha* in the vicinity of Port Aransas, Texas. *Contr. Mar. Sci.* **15**: 205–285.

KERMARREC, A. 1970. A propos d'une eventuelle parente de deux Chlorophycees marine *Acrochaete repens* et *Bolbocoleon piliferum* (Chaetophoraceae – Ulotrichales). *Cah. Biol. Mar.* **11**: 485–490.

KERMARREC, M. A. 1980. Sur la phase de la meiose dans le cycle de deux Chlorophycees marines: *Bryopsis plumosa* (Huds.) C. Ag. et *Bryopsis hypnoides* Lamouroux (Codiales). *Cah. Biol. Mar.* **23**: 443–466.

KJELLMAN, F. R. 1877. Om Spetsbergens marina klorofyllförande Thallophyter, II. *Bih. K. Svenska VetenskAkad. Handl.* **4**(6): 1–61.

KJELLMAN, F. R. 1883. The algae of the Arctic Sea. *K. Svenska VetenskAkad. Handl.* **20**(5): 1–351.

KJELLMAN, F. R. 1893. Studier öfver Chlorophyceslägtet *Acrosiphonia* J. G. Ag. och dess Skandinaviska arter. *Bih. K. Svenska VetenskAkad. Handl.* **18** Afd3(5): 1–114.

KLEBS, G. 1883. Über die Organisation einiger Flagellaten-Gruppen und ihre Beziehungen zu Algen und Infusorien. *Unters. Bot. Inst. Tübingen* **1**: 233–362.

KLUGH, A. B. 1925. Ecological polymorphism in *Enteromorpha crinita*. *Rhodora* **24**: 50–55.

KNEBEL, G. 1935. Monographie der Algenreihe der Prasiolales, insbesondere von *Prasiola crispa*. *Hedwigia* **75**: 1–120.

KNIGHT, M. & PARKE, M. 1931. Manx algae. An algal survey of the south end of the Isle of Man. *L.M.B.C. Mem. Typ. Br. Mar. Pl. Anim.* **30**: 1–155.

KOBARA, T. & CHIHARA, M. 1981. Laboratory culture and taxonomy of two species of *Derbesia* (Class Chlorophyceae) in Japan. *Bot. Mag. Tokyo* **94**: 1–10.

KOEMAN, R. P. T. & HOEK, C. VAN DEN 1981. The taxonomy of *Ulva* (Chlorophyceae) in the Netherlands. *Br. Phycol. J.* **16**: 9–53.

KÖHLER, K. 1956. Entwicklungsgeschichte, Geschlechtsbestimung und Befruchtung bei *Chaetomorpha*. *Arch. Protistenk.* **101**: 223–268.

KORNMANN, P. 1938. Zur Entwicklungsgeschichte von *Derbesia* und *Halicystis*. *Planta* **28**: 464–470.

KORNMANN, P. 1956. Zur Morphologie und Entwicklung von *Percursaria percursa*. *Helgoländer wiss. Meeresunters.* **5**: 259–272.

KORNMANN, P. 1959. Die heterogene Gattung *Gomontia*. I. Der sporangiale Anteil, *Codiolum polyrhizum*. *Helgoländer wiss. Meeresunters.* **6**: 229–238.

KORNMANN, P. 1959. *Die Entwicklung von Entocladia Wittrockii. Zusammenfassung eines auf der Jahrestagung der Studiengesellschaft zur Erforschung von Meeresalgen.* Hamburg [mimeographed].

KORNMANN, P. 1960. Die heterogene Gattung *Gomontia.* II. Der fädige Anteil, *Eugomontia sacculata* nov. gen. nov. spec. *Helgoländer wiss. Meeresunters.* 7: 59–71.

KORNMANN, P. 1961a. Über *Spongomorpha lanosa* und ihre Sporophytenformen. *Helgoländer wiss. Meeresunters.* 7: 195–205.

KORNMANN, P. 1961b. Über *Codiolum* und *Urospora. Helgoländer wiss. Meeresunters.* 8: 42–57.

KORNMANN, P. 1962. Die Entwicklung von *Monostroma grevillei. Helgoländer wiss. Meeresunters.* 8: 195–202.

KORNMANN, P. 1964. Die *Ulothrix*-Arten von Helgoland. I. *Helgoländer wiss. Meeresunters.* 11: 27–38.

KORNMANN, P. 1965. Was ist *Acrosiphonia arcta* ?. *Helgoländer wiss. Meeresunters.* 12: 40–51.

KORNMANN, P. 1966a. *Hormiscia* neu definiert. *Helgoländer wiss. Meeresunters.* 13: 408–425.

KORNMANN, P. 1966b. Eine erbliche variante von *Derbesia marina. Naturwissenschaften* 53: 161.

KORNMANN, P. 1969. Characterization of the *Chaetomorpha* species of Helgoland and List/Sylt. *Proc. Int. Seaweed Symp.* 6: 223–224.

KORNMANN, P. 1970a. Eine Mutation bei der siphonalen Grünalge *Derbesia marina. Helgoländer wiss. Meeresunters.* 21: 1–8.

KORNMANN, P. 1970b. Phylogenetische Beziehungen in der Grünalgengattung *Acrosiphonia. Helgoländer wiss. Meeresunters.* 21: 292–304.

KORNMANN, P. 1972. Ein Beitrag zur Taxonomie der Gattung *Chaetomorpha. Helgoländer wiss. Meeresunters.* 23: 1–31.

KORNMANN, P. 1973. Codiolophyceae, a new class of Chlorophyta. *Helgoländer wiss. Meeresunters.* 25: 1–31.

KORNMANN, P. & SAHLING, P. H. 1962. Zur Taxonomie und Entwicklung der *Monostroma*-Arten von Helgoland. *Helgoländer wiss. Meeresunters.* 8: 302–320.

KORNMANN, P. & SAHLING, P. H. 1974. Prasiolales (Chlorophyta) von Helgoland. *Helgoländer wiss. Meeresunters.* 26: 99–133.

KORNMANN, P. & SAHLING, P. H. 1976. Wiedereinführung von *Bryopsis lyngbyei* (Bryopsidales, Chlorophyta) als selbständige Art. *Helgoländer wiss. Meeresunters.* 28: 217–225.

KORNMANN, P. & SAHLING, P. H. 1977. Meeresalgen von Helgoland. Benthische Grün-, Braun- und Rotalgen. *Helgoländer wiss. Meeresunters.* 29: 1–289.

KORNMANN, P. & SAHLING, P. H. 1978. Die *Blidingia*-Arten von Helgoland (Ulvales, Chlorophyta). *Helgoländer wiss. Meeresunters.* 31: 391–413.

KORNMANN, P. & SAHLING, P. H. 1980. *Ostreobium quekettii* (Codiales, Chlorophyta). *Helgoländer wiss. Meeresunters.* 34: 115–122.

KORNMANN, P. & SAHLING, P. H. 1983. Meeresalgen von Helgoland: Ergänzung. *Helgoländer wiss. Meeresunters.* 36: 1–65.

KORSHIKOV, A. A. 1926. On some new organisms from the groups of Volvocales and Protococccales, and on the genetic relations of these groups. *Arch. Protistenk.* 55: 339–503.

KOSTER, J. TH. 1955. The genus *Rhizoclonium* in the Netherlands. *Pubbl. Staz. Zool. Napoli* 27: 335–357.

KRISTIANSEN, A. 1972. A seasonal study of the marine algal vegetation of Tuborg Harbour, The Sound, Denmark. *Bot. Tidsskr.* 67: 201–244.

KUCKUCK, P. 1894. Bemerkungen zur marinen Algen-vegetation von Helgoland. *Wiss. Meeresunters. (Helgoland) N.F.* 1(1): 223–263.

KUNIEDA, H. 1934. On the life history of *Monostroma. Proc. Imp. Acad. Tokyo* 10: 103–106.

KÜTZING, F. T. 1833. Algologische Mittheilungen. *Flora, Jena* 16: 511–521.

KÜTZING, F. T. 1843a. *Phycologia Generalis.* Leipzig.

KÜTZING, F. T. 1843b. Ueber die Systematische Eintheilung der Algen. *Linnaea* 17: 75–107.

KÜTZING, F. T. 1845. *Phycologia Germanica.* Nordhausen.

KÜTZING, F. T. 1847. Diagnosen und Bemerkungen zu neuen oder kritischen Algen. *Bot. Zeit.* 5: 164–167, 177–180.

KÜTZING, F. T. 1849. *Species Algarum.* Lipsiae.

KÜTZING, F. T. 1856. *Tabulae Phycologicae* **6**. Nordhausen.

KYLIN, H. 1910. Zur Kenntnis der Algenflora der Norwegischen Westküste. *Ark. Bot.* **10**(1): 1–37.

KYLIN, H. 1935. Über einige kalkbohrende Chlorophyceen. *K. Fysiogr. Sällsk. Lund. Förh.* **5**: 186–204.

KYLIN, H. 1938. Über die Chlorophyceengattungen *Entocladia* und *Ectochaete. Bot. Notiser* **1938**: 67–76.

KYLIN, H. 1947. Über die Fortpflanzungsverhältnisse in der Ordnung Ulvales. *K. Fysiogr. Sällsk. Lund. Förh* **17**: 174–182.

KYLIN, H. 1949. Die Chlorophyceen der Schwedischen Westküste. *Acta Univ. lund.* N.F, Avd. 2, **45**(4): 1–79.

LAGERHEIM, G. 1883. Bidrag till Sveriges Algflora. *Ofvers. K. VetenskAkad. Förh. Stockh.* **40**(1/2): 37–78.

LAGERHEIM, G. 1884. Om *Chlorochytrium cohnii* Wright och dess föhållende till nästående arter. *Ofvers. K. VetenskAkad. Förh. Stockh.* **41**(7): 91–97.

LAGERHEIM, G. 1885. *Codiolum polyrhizum* n.sp. Ett bidrag till Kännedomen om slägtet *Codiolum* A.Br. *Ofvers. K. VetenskAkad. Förh. Stockh.* **42**(8): 21–31.

LAKOWITZ, K. 1929. *Die Algenflora der gesamten Östsee*. Danzig.

LAMI, R. 1935. Le genre *Ulvella* Crn. dans la région Malouine. *Archs Mus. Natn. Hist. Nat. Paris* Sér. 6, **12**: 555–558.

LAMOUROUX, M. J. V. F. 1809. Mémoire sur trois nouveaux genres de la famille des algues marines. *J. Bot., Paris* **2**: 129–135.

LAMOUROUX, M. J. V. F. Essai sur les genres de las famille des thalassiophytes non articulées. *Annls Mus. Hist. Nat., Paris* **20**: 267–293.

LE JOLIS, A. 1863. *Liste des Algues Marines de Cherbourg*. Paris & Cherbourg.

LERSTON, N. R. & VOTH, P. D. 1960. Experimental control of zooid discharge and rhizoid formation in the green alga *Enteromorpha. Bot. Gaz.* **122**: 33–45.

LEVAN, A. & LEVRING, T. 1942. Some experiments on c-mitotic relations within Chlorophyceae and Phaeophyceae. *Hereditas* **28**: 400–408.

LEVRING, T. 1937. Zur Kenntnis der Algenflora der Norwegischen Westküste. *Acta Univ. lund.* N.F, Avd. 2, **33**(8): 1–148.

LEWIS, J. R. 1964. *The Ecology of Rocky Shores*. London.

LIGHTFOOT, J. 1777. *Flora Scotica* **2**. London.

LINDENBERG, 1840. *Conferva lehmanniana* n.sp. *Linnaea* **14**: 179–180.

LINDLEY, J. 1843. *Tetranema mexicanum. Bot. Reg.* **29**:1–85

LINK, H.F. 1820. Epistola de algis aquaticis in genera disponendis. *In*, Nees von Esenbeck, C. D. (Ed.), *Horae Physicae Berolinensis*. Bonnae.

LINNAEUS, C. 1753. *Species Plantarum* **2**. Holmiae.

LINNAEUS, C. 1758. *Systema Naturae* **1** ed. 2. Holmiae.

LOBBAN, C. S. & WYNNE, M. J. 1981. *The Biology of Seaweeds*. Botanical Monographs **17**. Oxford.

LOKHORST, G. M. 1978. Taxonomic studies on the marine and brackish-water species of *Ulothrix* (Ulotrichales, Chlorophyceae) in western Europe. *Blumea* **24**: 191–299.

LOKHORST, G. M. & TRASK, B. J. 1981. Taxonomic studies on *Urospora* (Acrosiphoniales, Chlorophyceae) in western Europe. *Acta Bot. Neerl.* **30**: 353–431.

LOKHORST, G. M. & VROMAN, M. 1974. Taxonomic studies on the genus *Ulothrix* (Ulotrichales, Chlorophyceae) III. *Acta Bot. Neerl.* **23**: 561–602.

LORENZ, J. R. 1855. Die Stratonomie von *Aegagropila sauteri. Denkschr. Akad. Wiss. Wien* **10**: 147–172.

LUND, S. 1959. The marine algae of East Greenland I. Taxonomical part. *Meddr Grønland* **156**: 1–247.

LYNGBYE, H. C. 1819. *Tentamen Hydrophytologiae Danicae*. Hafniae.

MAGGS, C. A., FREAMHAINN, M. T. & GUIRY, M. D. 1983. A study of the marine algae of subtidal cliffs in Loch Hyne (Ine), Co. Cork. *Proc. R. Ir. Acad.* **83B**: 251–266.

MAGGS, C. A. & GUIRY, M. D. 1981. Le *Cladophora pygmaea* Reinke, espèce nouvelle pour les côtes du France. *Trav. Stn Biol. Roscoff* **27**: 11–13.

MARCHAND, L. 1895. *Synopsis et Tableau Synoptique des Familles qui Compose la Classe des Phycophytes (Algues, Diatomée et Bactériene)*. Paris.

MARCHEWIANKA, M. 1924. Z flory glonow polskiego Baltyku. (Beiträge zu Algenflora der Ostsee). *Spraw. Kom. Fizjogr. Krakow* **58/59**: 33–45.

MASSEE, G. 1885. Marine algae of the Scarborough district. *Naturalist, Hull* N.S. **10**: 300–306.

MATHIESON, A. C. & BURNS, R. L. The discovery of *Halicystis ovalis* (Lyngbye) Areschoug in New England. *J. Phycol.* **6**: 404–405.

MARTIUS, C. F. P. 1833. *Flora Brasiliensis* **1**. Stuttgartiae et Tubingiae.

MATTOX, K. R. & STEWART, K. D. 1985. Classification of the green algae: a concept based on comparative cytology, pp. 29–72. *In*, Irvine, D. E. G. & John, D. M. (Eds), *Systematics of the Green Algae*. Systematics Association Special Volume **27**. London.

MAYHOUB, H. 1974a. Observations et explications au sujet de l'ecologie et de la répartition des *Pseudobryopsis myura* (Ag.) Berthold (Codiales, Bryopsidacées). *Bull. Soc. Phycol. Fr.* **19**: 164–167.

MAYHOUB, H. 1974b. Reproduction sexuée et cycle du développement de *Pseudobryopsis myura* (Ag.) Berthold (Chlorophycée, Codiale). *C. r. hebd. Séanc. Acad. Sci., Paris* Ser. D, **278**: 867–870.

MELKONIAN, M. 1985. Flageller apparatus ultrastructure in relation to green algal classification, pp. 73–120. *In*, Irvine, D. E. G. & John, D. M. (Eds), *Systematics of the Green Algae*. Systematics Association Special Volume **27**. London.

MENEGHINI, G. 1838. Cenni sulla organografia e fisiologia delle alghe. *Nuovi Saggi Accad., Padova* **4**: 325–328.

MENEGHINI, G. 1842. Monographia Nostochinearum Italiarum addito specimine de rivularis. *Memorie Accad. Sci. Torino* Ser2, **5**: 1–143.

MIGITA, S. 1967. Life cycle of *Capsosiphon fulvescens* (C. Agardh) Setchell and Gardner. *Bull. Fac. Fish. Nagasaki Univ.* **22**: 21–31.

MOESTRUP, O. 1969. Observations on *Bolbocoleon piliferum*. Formation of hairs, reproduction and chromosome number. *Bot. Tidsskr.* **64**: 169–175.

MOESTRUP, O. 1978. On the phylogenetic validity of the flagellar apparatus in green algae and other chlorophyll *a* and *b* containing plants. *Biosystems* **10**: 117–144.

MOORE, G. T. 1900. New or little known unicellular algae I. *Chlorocystis cohnii. Bot. Gaz.* **30**: 100–112.

MONTAGNE, C. 1849. Sixième centurie de plantes cellulaires nouvelles, faut indigènes qu'exotiques. *Ann. Sci. nat. Bot.* Sér. 3, **11**: 33–66.

MORIS, J. & DE NOTARIS, J. 1839. Florula Caprariae. *Memorie Accad. Sci. Torino* Ser. 2, **2**: 59–300.

MORTON, O. 1978. Some interesting records of algae from Ireland. *Irish Nat. J.* **19**: 240–242.

MOSS, B. & MARSLAND, A. 1976. Regeneration of *Enteromorpha*. *Br. Phycol. J.* **11**: 309–313.

MÜLLER, O. F. 1778. *Flora Danica* **5**, Fasc. 13. Havniae.

MÜLLER–STÖLL, W. R. 1952. Über Regeneration und Polarität bei *Enteromorpha*. *Flora* **139**: 148–180.

NÄGELI, C. 1849. Gattungen einzelliger Algen physiologisch und systematisch bearbeitet. *Neue Denkschr. Allg. Schweiz. Ges. Ges. Naturw.* **10**(7): 1–139.

NÄGELI, C. 1847. Die neuen Algensysteme und Versuch zur Begründing eines eigenen Systems der Algen und Florideen. *Neue Denkschr. Allg. Schweiz. Ges. Ges. Naturw.* **9**(2): 1–275.

NEUMANN, K. 1969a. Protonema mit Riesenkern bei der siphonalen Grünalge *Bryopsis hypnoides* und weitere cytologische Befunde. *Helgoländer wiss. Meeresunters.* **19**: 45–57.

NEUMANN, K. 1969b. Beitrag zur Cytologie und Entwicklung der siphonalen Grünalge *Derbesia marina*. *Helgoländer wiss. Meeresunters.* **19**: 355–375.

NEUMANN, K. 1974. Zur Entwicklungsgeschichte und systematik der siphonalen Grünalgen *Derbesia* und *Bryopsis. Botanica mar.* **17**: 176–185.

NEWTON, L. 1931. *A Handbook of the British Seaweeds*. London.

NICOLAI, E. & PRESTON, R. D. 1952. Cell wall studies in the Chlorophyceae. I. A general survey of submicroscopic structure in filamentous species. *Proc. R. Soc.* Ser. B, **140**: 244–274.

NICOLAI, E. & PRESTON, R. D. 1959. Cell wall studies on the Chlorophyceae. III. Differences in structure and development in the Cladophoraceae. *Proc. R. Soc. Ser.* B, **151**: 244–255.

NIELSEN, R. 1972. A study of the shell-boring marine algae around the Danish Island Laesø. *Bot. Tidsskr.* **67**: 245–269.

NIELSEN, R. 1977. Culture studies on *Ulvella lens* and *Ulvella setchellii. Br. Phycol. J.* **12**: 1–5.

NIELSEN, R. 1978. Variation in *Ochlochaete hystrix* (Chaetophorales, Chlorophyceae) studied in culture. *J. Phycol.* **14**: 127–131.

NIELSEN, R. 1979. Culture studies on the type species of *Acrochaete, Bolbocoleon* and *Entocladia* (Chaetophoraceae, Chlorophyceae). *Bot. Notiser* **132**: 441–449.

NIELSEN, R. 1980. A comparative study of five marine Chaetophoraceae. *Br. Phycol J.* **15**: 131–138.

NIELSEN, R. 1983. Culture studies of *Acrochaete leptochaete* comb. nov. and *A. wittrockii* comb. nov. (Chaetophoraceae, Chlorophyceae). *Nord. J. Bot.* **3**: 689–694.

NIELSEN, R. & PEDERSEN, P. M. 1977. Separation of *Syncoryne reinkei* nov. gen. nov. sp. from *Pringsheimiella scutata* (Chlorophyceae, Chaetophoraceae). *Phycologia* **16**: 411–416.

NIENHUIS, P. H. 1975. *Biosystematics and Ecology of* Rhizoclonium riparium *(Roth) Harv. (Chlorophyceae: Cladophorales) in the Estuarine Area of the Rivers Rhine, Meuse and Scheldt.* Rotterdam.

NIIZEKI, S. 1957. Cytological study of swarmer formation in *Enteromorpha linza. Nat. Sci. Rep. Ochanamizu Univ.* **8**: 45–51.

NISHIMURA, M. & KANND, R. 1927. On the asexual reproduction of *Aegagropila sauteri* (Nees) Kütz. *Bot. Mag. Tokyo* **41**: 432–438.

NORTON, T. A. (Ed.) 1985. *Provisional Atlas of the Marine Algae of Britain and Ireland.* NERC, Huntingdon.

NORTON, T. A. 1987. Elsie M. Burrows (1913–1986). *Br. Phycol. J.* **22**: 317–319.

NORTON, T. A., MCALLISTER, H. A., CONWAY, E. & IRVINE, L. M. 1969. The marine algae of the Hebridean Island of Colonsay. *Br. Phycol. J.* **4**: 125–136.

NOTARIS, G. DE 1846. *Prospetto della Flora Ligustica e dei Zoofiti del Mare Ligustico.* Genova.

O'DONNELL, J. J. & PERCIVAL, E. 1959. Structural investigation on the water-soluble polysaccharides from the green sea-weed *Acrosiphonia centralis (Spongomorpha arcta). J. Chem. Soc.* **1959**: 2168–2178.

OHGAI, M., FUJIYAMA, T. & MATSUI, T. 1975. Studies on the life-history of marine *Ulothrix*. I. On the asexual reproduction of *Ulothrix implexa* Kuetzing prox. *J. Shimonoseki Univ. Fish.* **23**: 137–144.

OHGAI, M. & FUJIYAMA, T. 1976. Studies on the life-history of *Ulothrix*. II. On the sexual reproduction and life-history of *Ulothrix implexa* Kuetzing prox. *J. Shiminoseki Univ. Fish.* **24**: 337–342.

O'KELLY, C. J. 1980. *The Taxonomy and Systematics of Certain Marine Algae Referable to the Chaetophoraceae and Chaetosiphonaceae (Chlorophyta).* Thesis, University of Washington.

O'KELLY, C. J. & FLOYD, G. L. 1985. Correlations among patterns of sporangial structure and development, life-histories and ultrastructural features in the Ulvophyceae. pp. 121–156. *In*, Irvine, D. E. G. & John, D. M. (Eds), *Systematics of the Green Algae*. Systematics Association Special Volume **27**. London.

O'KELLY, C. J. & YARISH, C. 1980. Observations on marine Chaetophoraceae (Chlorophyta). I. Sporangial ontogeny in the type species of *Entocladia* and *Phaeophila. J. Phycol.* **16**: 549–558.

O'KELLY, C. J. & YARISH, C. 1981. Observations on marine Chaetophoraceae (Chlorophyta). II. On the circumscription of the genus *Entocladia* Reinke. *Phycologia* **20**: 32–45.

OLIVI, G. 1792. *Zoologia Adriatica. . . .* Bassano.

OLTMANNS, F. 1894. Ueber einige parasitische Meeresalgen. *Bot. Zeit.* **52**: 207–215.

OLTMANNS, F. 1904. *Morphologie und Biologie der Algen* **1**. Jena.

PAGE, J. Z. 1970. Existence of a *Derbesia* phase in the life-history of *Halicystis osterhoutii* Blinks and Blinks. *J. Phycol.* **6**: 375–380.

PAGE, J. Z. & KINGSBURY, J. M. 1968. Culture studies on the marine green alga *Halicystis parvula – Derbesia tenuissima.* II. Synchrony and periodicity in gamete formation. *Am. J. Bot.* **55**: 1–11.

PAGE, J. Z. & SWEENY, B. M. 1969. Culture studies on the marine green alga *Halicystis parvula –*

Derbesia tenuissima. III. Control of gamete formation by an endogenous rhythm. *J. Phycol.* **4**: 253–260.

PAPENFUSS, G. F. 1960. On the genera of the Ulvales and the status of the order. *J. Linn. Soc. Bot.* **56**: 303–318.

PAPENFUSS, G. F. 1962. On the circumscription of the green algal genera *Ulvella* and *Pilinia. Phykos.* **1**: 6–12.

PARKE, M. 1935. Algological records for the Manx region. *Rep. Mar. Biol. Stn Port Erin* **48**: 29–32.

PARKE, M. & DIXON, P. S. 1968. Check-list of British marine algae – second revision. *J. Mar. Biol. Ass. U.K.* **48**: 783–832.

PARKE, M. & DIXON, P. S. 1976. Check-list of British marine algae – third revision. *J. mar. biol. Ass. U.K.* **56**: 527–594.

PARKER, B. Significance of cell wall chemistry to phylogeny in the algae. *Ann. N.Y. Acad. Sci.* **175**: 417–428.

PARKES, H. M. 1975. Records of *Codium* species in Ireland. *Proc. R. Ir. Acad.* **75B**: 125–134.

PASCHER, A. 1915. *Die Süsswasser–Flora Deutschlands, Österreichs und der Schweiz.* **5**: Chlorophyceae II. Jena.

PEDERSEN, P. M. 1973. Preliminary note on some marine algae from South Greenland. *Bot. Tidsskr.* **68**: 145–149.

PERROT, Y. 1969. Sur le cycle ontogénétique et chromosomique du *Pseudopringsheimia confluens* (Rosenv.) Wille. *C. r. hebd. Séanc. Acad. Sci., Paris* Sér. D, **268**: 279–282.

PERROT, Y. 1965. Sur le cycle de reproduction d'une Cladophoracée marine de la région de Roscoff: *Lola implexa* (Harvey). *C. r. hebd. Séanc. Acad. Sci., Paris* **261**: 503–506.

PERROT, Y. 1972. Les *Ulothrix* marins de Roscoff et le problème de leur cycle de reproduction. *Mém. Soc. Bot. Fr.* **1972**: 67–74.

PIERRE, J. B. L. 1890. *Notes Botaniques. Sapotacées.* Paris.

PICKETT-HEAPS, J. D. 1975. *Green Algae. Structure. Reproduction and Evolution in Selected Genera.* Sunderland.

PICKETT-HEAPS, J. D. & MARCHANT, H. J. 1972. The phylogeny of the green algae: a new proposal. *Cytobios* **6**: 255–264.

POLDERMAN, P. J. G. 1975. Some notes on the algal vegetation of two brackish polders on Texel (The Netherlands). *Hydrobiol. Bull.* **9**: 23–34.

POLDERMAN, P. J. G. 1976. *Wittrockiella paradoxa* Wille (Cladophoraceae) in N.W. European saltmarshes. *Hydrobiol. Bull.* **10**: 98–103.

POLDERMAN, P. J. G. 1978. Algae of the saltmarshes on the south and southwest coasts of England. *Br. Phycol. J.* **13**: 235–240.

POSTELS, A. & RUPRECHT, F. 1840. *Illustrationes Algarum.* Petropoli.

PRESL, K. B. 1851. Epimeliae Botanicae. *Abh. K. Böhm. Ges. Wiss.* Ser. 5, **6**: 361–624.

PRICE, J. H., HEPTON, C. E. L. & HONEY, S. I. 1980. The inshore benthic biota of the Lizard Peninsula, south west Cornwall. 1. The marine algae: history; Chlorophyta; Phaeophyta. *Cornish Stud.* **7**: 7–37.

PRICE, W. M. 1967. *Some Aspects of the Biology and Taxonomy of the Unbranched Cladophorales.* Thesis, University of Liverpool.

PRINGLE, 1975. *Studies on Variation in Three Taxa of* Enteromorpha. Ph.D. Thesis, Dalhousie University, Halifax, N.S.

PRINGSHEIM, N. 1863. Beiträge zur Morphologie der Meeres-Algen. *Abh. Akad. Wiss., Berlin* **1862**: 1–37.

PRINTZ, H. 1926. Die Algenvegetation des Trondhjemsfjordes. *Skr. Norske Vidensk-Akad. Mat. -Nat. Kl.* **1**(5): 1–266.

PRINTZ, H. 1927. Chlorophyceae, pp. 1–411. *In*, Engler, A. & Prantl, K. (Eds), *Die natürlichen Pflanzenfamilien* Ed. 3. Leipzig.

RABENHORST, L. 1864. *Flora Europaea Algarum Aquae Dulcis et Submarinae* Sectio I. Lipsiae.

RABENHORST, L. 1868. *Flora Europaea Algarum Aquae Dulcis et Submarinae* Sectio III. Lipsiae.

RAFINESQUE, C. S. 1810. *Caratteri di Alcuni Nuovi Generi e Nuove Specie di Animale e Plante della Sicilia*. Palermo.

RAMANATHAN, K. R. 1939. The morphology, cytology, and alternation of generations in *Enteromorpha compressa* (L.) Grev. var. *lingulata* (J.Ag.) Hauck. *Ann. Bot.* N.S. **3**: 375–398.

REED, R. H. & RUSSELL, G. 1978. Salinity fluctuations and their influence on 'Bottle Brush' morphogenesis in *Enteromorpha intestinalis* (L.) Link. *Br. Phycol. J.* **13**: 149–153.

REES, A. A. & RUSSELL, G. 1977. The benthic phase of *Chlorococcum submarinum* Ålvik. *Br. Phycol. J.* **12**: 122.

REINBOLD, T. 1893. Revision von Jürgens Algae Aquaticae. I. Die Algen des Meeres und des Brackwassers. *Nuova Notarisia* Ser. 4: 192–206.

REINHARDT, L. 1885. Material zur Morphologie und Systematik der Algen des Schwarzen Meeres. *Denkschr. Neu-Russ. Ges. Naturforscher.* **9**: 199–510.

REINKE, J. 1879. Zwei parasitische Algen. *Bot. Zeit* **37**: 472–478.

REINKE, J. 1888. Einige neue braune und grüne Algen der Kieler Bucht. *Ber. dt. Bot. ges.* **6**: 240–241.

REINKE, J. 1889. Algenflora der westlichen Ostsee. Deutscher Antheils. VI *Ber. Com. wiss. Untersuch. dt. Meere in Kiel* Jg **17–21**, Heft **1**: 1–101.

REINKE, J. 1889a. *Atlas deutscher Meeresalgen*. I. Berlin.

REINSCH, J. 1879. Ein neues Genus der Chroolepideae. *Bot. Zeit.* **37**: 361–366.

RIETEMA, H. 1969. A new type of life history in *Bryopsis* (Chlorophyceae, Caulerpales). *Acta Bot. Neerl.* **18**: 615–619.

RIETEMA, H. 1970. Life-histories of *Bryopsis plumosa* (Chlorophyceae, Caulerpales) from European coasts. *Acta Bot. Neerl.* **19**: 859–866.

RIETEMA, H. 1971a. Life history-studies in the genus *Bryopsis* (Chlorophyceae). III. The life-history of *Bryopsis monoica* Funk. *Acta Bot. Neerl.* **20**: 205–210.

RIETEMA, H. 1971b. Life-history studies in the genus *Bryopsis* (Chlorophyceae). IV. Life-histories in *Bryopsis hypnoides* Lamx. from different points along the European coasts. *Acta Bot. Neerl.* **20**: 291–298.

RIETEMA, H. 1975. *Comparative Investigations on the Life-histories and Reproduction of Some Species in the Siphoneous Green Algal Genera Bryopsis and Derbesia*. Groningen.

ROSENVINGE, L. K. 1892. Om nogle vaextforheld hos slaegterne *Cladophora* og *Chaetomorpha*. *Bot. Tidsskr.* **18**: 29–64.

ROSENVINGE, L. K. 1893. Grønlands Havalger. *Meddr Grønland* **3**: 763–981.

ROTH, A. G. 1797. *Catalecta Botanica* Fasc. 1. Lipsiae.

ROTH, A. G. 1800. *Catalecta Botanica* Fasc. 2. Lipsiae.

ROTH, A. G. 1806. *Catalecta Botanica* Fasc. 3. Lipsiae.

ROUND, F. E. 1971. The taxonomy of the Chlorophyta II. *Br. Phycol. J.* **6**: 235–264.

ROUSSEL, H. F. A. 1806. *Flora du Calvados et des Terrains Adjacens* Ed. 2. Caen.

RUPRECHT, F. J. 1851. Tange des Ochotskischen Meeres. *In*, Middendorf, A. von (Ed.), *Reise in den aüssersten Norden und Östen Sibiriens* 1(2). St Petersburg.

RUSSELL, G. 1973. The Phaeophyta: a synopsis of some recent developments. *Oceanogr. Mar. Biol. Ann. Rev.* **11**: 45–88.

SCAGEL, R. F. 1957. An annotated list of the marine algae of British Columbia and Washington (including keys to genera). *Bull. Natn. Mus. Can.* **150**: 1–289.

SCAGEL, R. F. 1966. Marine algae of British Columbia and Northern Washington, Part 1: Chlorophyceae (green algae). *Bull. Natn. Mus. Can.* **207**: viii + 1–257.

SCHMIDLE, W. 1899. Algologische notizen. *Allg. bot. Z.* **5**: 57–58.

SCHMIDT, 1893. *In*, Murray G. (Ed.), On *Halicystis* and *Valonia*. *Phycol. mem.* **2**: 47–52.

SCHMIDT, O. C. 1923. Beiträge zur Kenntnis der Gattung *Codium* Stackh. *Biblthca Bot.* **91**: 1–68.

SCHULZER, S. VON MÜGGENBERG, KANITZ, A. & KNAPP, J. A. 1866. Die bisher bekannten Pflanzen Slavoniens. *Verh. Zool.–Bot. Ges. Wien* **16** (Abhandlung): 1–172.

SCHUSSNIG, B. 1932. Der Generations- und Phasenwechsel bei den Chlorophyceen. *Öst. bot. Z.* **81**: 296–298.

SCHUSSNIG, B. 1939. Zur heterochromosome-Frage bei der Gattung *Cladophora*. *Öst. bot. Z.* **88**: 210–217.

SCHUSSNIG, B. 1960. Systematik und Phylogenie der Algen. Chaetophorales. *Forschr. Bot., Berl.* **22**: 57.

SCHUSSNIG, E. 1960. *Handbuch der Protophytenkunde* 2. Jena.

SEARS, J. R. 1967. Mitotic waves in the green alga *Blastophysa rhizopus* as related to coenocyte form. *J. Phycol.* **3**: 136–139.

SEARS, J. R. 1968. Developmental morphology and systematics of the siphonaceous green alga *Blastophysa rhizopus. J. Phycol.* 3(suppl.): 3.

SEARS, J. R. & WILCE, R. T. 1970. Reproduction and systematics of the marine alga *Derbesia* (Chlorophyceae) in New England. *J. Phycol.* **6**: 381–392.

SETCHELL, W. A. & GARDNER, N. L. 1920a. Phycological contributions I. *Univ. Calif. Publs. Bot.* **7**: 279–324.

SETCHELL, W. A. & GARDNER, N. L. 1920b. The marine algae of the Pacific coast of North America. II Chlorophyceae. *Univ. Calif. Publs. Bot.* **8**: 139–374.

SILVA, M. W. R. N. DE 1969. *An Experimental Approach to the Taxonomy of the Genus* Enteromorpha *(L.) Link.* Thesis, University of Liverpool.

SILVA, M. R. W. N. DE 1978. An experimental assessment of the variability of the character of the chloroplast as taxonomic [sic] criteria in the genus *Enteromorpha* (L.) Link, algae. *Malayan J. Sci.* **5** (A): 119–127.

SILVA, M. R. W. N. DE & BURROWS, E. M. 1973. An experimental assessment of the status of the species *Enteromorpha intestinalis* (L.) Link and *Enteromorpha compressa* (L.) Grev. *J. Mar. Biol. Ass. U.K.* **53**: 895–904.

SILVA, P. C. 1950. Generic names of algae proposed for conservation. *Hydrobiologia* **2**: 252–280.

SILVA, P. C. 1952. A review of nomenclatural conservation in the algae from the point of view of the type method. *Univ. Calif. Publs. Bot.* **25**: 241–323.

SILVA, P. C. 1955. The dichotomous species of *Codium* in Britain. *J. Mar. Biol. Ass. U.K.* **34**: 565–577.

SILVA, P. C. 1957. Notes on Pacific marine algae. *Madroño* **14**: 41–51.

SILVA, P. C. 1980. Names of classes and families of living algae. *Regnum Vegetabile* **103**: 1–156.

SILVA, P. C., MATTOX, K. R. & BLACKWELL, W. H. 1972. The generic name *Hormidium* as applied to green algae. *Taxon* **21**: 639–645.

SJÖSTEDT, G. 1922. Om *Prasiola cornucopiae* J. G. Ag. och *Prasiola stipitata* v. Suhr samt deras förhållende inbördes. *Bot. Notiser* 1922; 37–45.

SLUIMAN, H. J., STEWART, K. D. & MATTOX, K. R. 1980. The ultrastructure of the scaly zoospore of *Ulothrix zonata*: its taxonomic significance. *J. Phycol.* **16** (suppl.): 40.

SMITH, G. M. 1930. Observations on some siphoneous green algae of the Monterey Peninsula. *In, Contributions to Marine Biology. Lectures and Symposia Given at the Hopkins Marine Station Dec. 20–21, 1929.* Stanford.

SNOW, J. W. 1911. Two epiphytic algae. *Bot. Gaz.* **51**: 360–368.

SÖDERSTRÖM, J. 1963. Studies in *Cladophora. Botanica Gothoburg.* **1**: 1–147.

SOLIER, A. J. J. 1847. Mémoire sur deux algues zoosporées devant former un genre distinct, le genre *Derbesia. Ann. Sci. Nat. Bot.* Sér. 3, **7**: 157–166.

SOUTH, G. R. 1968. Aspects of the development and reproduction of *Acrochaete repens* and *Bolbocoleon piliferum. Can. J. Bot.* **46**: 101–113.

SOUTH, G. R. 1969. A study of *Bolbocoleon piliferum* Pringsh. *Proc. Int. Seaweed Symp.* **6**: 375–381.

SOUTH G. R. 1974. Contributions to the flora of the marine algae of eastern Canada, II. Family Chaetophoraceae. *Naturaliste Can.* **101**: 905–923.

SOUTH, G. R. 1976. A check-list of marine algae of eastern Canada – first revision. *J. Mar. Biol. Ass. U.K.* **56**: 817–843.

SOUTH, G. R. & CARDINAL, A. 1970. A checklist of marine algae of eastern Canada. *Can. J. Bot.* **48**: 2077–2095.

STACKHOUSE, J. 1795. *Nereis Britannica* Fasc. 1. Bathoniae.

STACKHOUSE, J. 1797. *Nereis Britannica* Fasc. 2. Bathoniae.

STEWART, K. D., MATTOX, K. R. & FLOYD, G. L. 1973. Mitosis, cytokinesis, the distribution of plasmodesmata, and other cytological characteristics in the Ulotrichales, Ulvales and Chaetophorales: phylogenetic and taxonomic considerations. *J. Phycol.* **9**: 128–141.

STEWART, K. D. & MATTOX, K. R. 1978. Structure and evolution in the flagellated cells of green algae and land plants. *Biosystems* **10**: 145–152.

STOCKMAYER, S. 1890. Ueber die Gattung *Rhizoclonium*. *Verh. Zool.- Bot. Ver. Wien* **40**: 571–586.

SUNDENE, O. 1953. The algal vegetation of the Oslofjord. *Skr. Norske Vidensk. -Akad. Mat. - Nat. Kl.* **1953**(2): 1–244,

SURINGAR, W. F. R. 1870. *Algae Japonicae Musei Botanicii Lugduno-Batavi.* Harlemi.

SVEDELIUS, N. 1901. *Studier öfver Östersjöns hafsalgflora.* Uppsala.

TATEWAKI, M. 1972. Life history and systematics in *Monostroma*. pp. 1–15. *In*, Abbot, I. A. & Kurogi, M. (Eds), *Contributions to the Systematics of Benthic Marine Algae of the North Pacific.* Kobe.

TATEWAKI, M. & NAGATA, K. 1970. Surviving protoplasts *in vitro* and their development in *Bryopsis*. *J. Phycol.* **6**: 401–403.

TAYLOR, W. R. 1937a. *Marine Algae of the Northeastern Coast of North America.* Ann Arbor.

TAYLOR, W. R. 1937b. Notes on north Atlantic marine algae. I. *Pap. Mich. Acad. Sci.* **22**: 225–233.

TAYLOR, W. R. 1957. *Marine Algae of the Northeastern Coast of North America.* Ed. 2. Ann Arbor.

THURET, G. 1854. Note sur la synonymie des *Ulva lactuca* et *latissima* L. suivie de quelques remarques sur la tribu des Ulvacées. *Mém. Soc. Sci. nat. Cherbourg.* **2**: 17–32.

TOKIDA, J. The marine algae of southern Saghalien. *Mem. Fac. Fish. Hokkaido Univ.* **2**: 1–263.

TREVISAN, V. B. A. 1842. *Prospetto della Flora Euganea.* Padova.

VALET, G. 1961. Les *Chaetomorpha* de la région de Montpellier-Sète. *Naturalia monspel.* Ser. Bot. **12**: 81–88.

VAN GOOR, A. C. J. 1923. Die holländischen Meeresalgen (Rhodophyceae, Phaeophyceae und Chlorophyceae) insbesondere der Umgebung von Helder des Wattenmeeres und der Zuidersee. *Verh. K. Akad-Wet. Amst.* Sect. 2, **23**: 1–232.

VAN DEN HOEK, C. 1963. *Revision of the European Species of Cladophora.* Leiden.

VAN DEN HOEK, C. 1966. Taxonomic criteria in four Chlorophycean genera. *Nova Hedwigia* **10**: 367–386.

VAN DEN HOEK, C. 1980. Chlorophyta: morphology and classification, pp. 86–132. *In*, Lobban, C. S. & Wynne, M. J. (Eds), *The Biology of Seaweeds.* Oxford.

VAN DEN HOEK, C. 1982. *A Taxonomic Revision of the American Species of Cladophora (Cladophoraceae) in the North Atlantic Ocean and their Geographical Distribution.* Amsterdam, Oxford, New York.

VAN DEN HOEK, C., DUCKER, S. C. & WOMERSLEY, H. B. S. 1984. *Wittrockiella salina* Chapman (Cladophorales, Chlorophyceae), a mat and ball forming alga. *Phycologia* **23**: 39–46.

VAUCHER, J. P. 1803. *Histoire des Conferves d'eau douce.* Genève.

VINOGRADOVA, K. L. 1969. A contribution to the taxonomy of the order Ulvales (Chlorophyta). *Bot. Zh. SSSR* **54**: 1347–1355.

VOUK, V. 1936. Studien über Adriatische Codiaceen. *Acta adriat.* **1**(8): 1–47.

WAERN, M. 1952. Rocky-shore algae in the Öregrund Archipelago. *Acta phytogeogr. Suec.* **30**: xvi + 1–298.

WARMING, E. 1884. *Haandbog i den Systematiske Botanik.* Kjøbenhavn.

WEBER, F. & MOHR, D. M. H. 1804. *Naturhistorische Reise durch einen Theil Schwedens.* Göttingen.

WEBBER, E. E. 1978. Phycological studies from the Marine Science Institute, Nahant, Massachusetts. V. Chlorophycophyta. *Botanica mar.* **21**: 393–395.

WENT, F. A. F. C. 1889. Les modes de reproduction du *Codium tomentosum*. *Nederl. Kruidk. Archf.* Ser. 2, **5**: 440–444.

WEST, G. S. 1904. *A Treatise on the British Freshwater Algae.* Cambridge.

WEST, G. S. 1916. *Algae I.* Cambridge.

WEST, G. S. & FRITSCH, F. E. 1927. *A treatise on the British Freshwater Algae.* Cambridge.

WILCE, R. T. 1970. *Cladophora pygmaea* Reinke in North America. *J. Phycol.* **6**: 260–263.

WILKINSON, M. 1969. *The Taxonomy and Autecology of Some British Marine Shell-Boring Green Algae.* Thesis, University of Liverpool.

WILKINSON, M. 1975. The occurrence of shell-boring *Phaeophila* species in Britain. *Br. Phycol. J.* **10**: 235–240.

WILKINSON, M. & BURROWS, E. M. 1972. An experimental taxonomic study of the algae confused under the name *Gomontia polyrhiza*. *J. Mar. Biol. Ass. U.K.* **52**: 49–57.

WILLE, N. 1880. Om en ny endophytisk alge. *Forh. VidenskSelsk. Krist.* **1880**: 1–4.

WILLE, N. 1890. Chlorophyceae, pp. 24–161. *In*, Engler, A. & Prantl, K. *Die Natürlichen Pflanzenfamilien.* **1**(2).

WILLE, N. 1901. *Studien über Chlorophyceen I–VII.* Christiana.

WILLE, N. 1906. Algologische Untersuchungen an der Biologischen Station in Drontheim I–VII I. Über die Entwicklung von *Prasiola furfuracea* (Fl. D.) Menegh. *K. Nor. Vidensk. Selsk. Skr.* **1906**: 1–38.

WILLE, N. 1909a. VII. Abteiling. Chlorophyceae, pp. 12–16. In Engler, A. (Ed.), *Syllabus der Pflanzenfamilien* 6 Afl. Berlin.

WILLE, N. 1909b. Algologische Notizen XV. Über *Wittrockiella* nov. gen. *Nytt Mag. Naturvid.* **47**: 209–225.

WILLE, N. 1909c. Conjugatae und Chlorophyceae. *In*, Engler, A. & Prantl, K. (Eds), *Die Natürlichen Pflanzenfamilien.* Nachtrag zum I. Teil, 2 Abteilung. Leipzig.

WILLIAMS, M. M. 1925. Contributions to the cytology and phylogeny of the siphonaceous algae. I. Cytology of the gametangia of *Codium tomentosum* (Stackh.). *Proc. Linn. Soc. N.S.W.* **50**: 98–111.

WITTROCK, V. B. 1866. *Försök till en Monograph Öfver Algslägtet* Monostroma. Stockholm.

WITTROCK, V. B. 1872. Om Gotlands och Ölands sötvattens-alger. *Bih. K. Svenska VetenskAkad. Handl.* **1**(1): 1–72.

WITTROCK, V. B. 1877. On the development and systematic arrangement of the Pithophoraceae. A new order of algae. *Nova Acta R. Soc. Scient. Upsal.* Ser. 3, Extra volume **1877**: 1–80.

WOMERSLEY, H. B. S. 1956. A critical survey of the marine algae of southern Australia. I. Chlorophyta. *Aust. J. Mar. Freshwat. Res.* **7**: 343–383.

WOOD, H. C. 1872. A contribution to the history of the fresh-water algae of North America. *Smithson. Contr. Knowl.* **19**: viii + 1–262.

WRIGHT, E. P. 1877. On a new species of parasitic green alga belonging to the genus *Chlorochytrium* of Cohn. *Trans. R. Ir. Acad.* **26**: 355–368.

WRIGHT, E. P. 1881. On a new genus and species of unicellular algae, living on the filaments of *Rhizoclonium casparyi*. *Trans. R. Ir. Acad.* **28**: 27–30.

WULFEN, X. 1803. *Cryptogamia aquatica.* Lipsiae.

YAMADA, Y. & SAITO, E. 1938. On some culture experiments with the swarmers of certain species belonging to the Ulvaceae. *Sci Pap. Inst. Algol. Res. Hokkaido Univ.* **2**: 35–51.

YARISH, C. 1975. A cultural assessment of the taxonomic criteria of selected marine Chaetophoraceae (Chlorophyta). *Nova Hedwigia* **26**: 385–430.

YARISH, C. 1976. Polymorphism of selected marine Chaetophoraceae (Chlorophyta). *Br. Phycol. J.* **11**: 29–38.

YENDO, K. 1917. Notes on algae new to Japan. VII. *Bot. Mag. Tokyo* **36**: 183–207.

ZINNECKER, E. 1935. Beiträge zur Entwicklungsgeschichte der Protophyten. XII. Reduktionsteilung, Kernphasenwechsel und Geschlechtbestimmung bei *Bryopsis plumosa* (Huds.) Ag. *Öst. Bot. Z.* **84**: 53–72.

Plate 1. *Monostroma grevillei*. Habit of plant. Hackemdown Pt., Kingsgate, Thanet, Kent; April 1985; epizoic on hydroids associated with mussel debris, littoral zone. Bar= 10 mm

Plate 2.   *Spongomorpha arcta.* Habit of plant. Peveril Pt., Swanage, Dorset; April, 1985; littoral zone.   Bar= 10 mm

Plate 3. *Enteromorpha intestinalis*. Habit of plant. Long Nose Spit, Margate, Kent; August, 1984; pool, lower littoral zone. Bar= 20 mm.

Plate 4.   *Enteromorpha linza*. Habit of plant. Long Nose Spit, Margate, Kent; April, 1985; pool, littoral zone.

Plate 5. *Ulva lactuca*. Habit of plant. Long Nose Spit. Margate, Kent; April, 1985; lower littoral zone. Bar= 20 mm

Plate 6. *Cladophora hutchinsiae*. Habit of plant. Long Nose Spit, Margate; August, 1984; pool, littoral zone. Bar= 10 mm

Plate 7. *Cladophora rupestris*. Habit of plant. Long Nose Spit, Margate; August, 1984; vertical concrete wall, upper littoral zone. Bar= 10 mm

Plate 8.   *Bryopsis plumosa.* Habit of plant. Long Nose Spit, Margate; August, 1984; channel,
lower littoral zone. Bar= 10 mm

Plate 9.   *Codium fragile* subspecies *tomentosoides*. Habit of plant. Northney, Hayling Island, Hampshire; August, 1984; side of floating pontoon. Bar= 20 mm

TAXONOMIC INDEX

Orders and families are given in capitals, whilst genera, species and infraspecific taxa are given in roman type. Synonyms are given in *italic* type. All taxa accepted in the present work and their main page numbers are given in **bold** type.

SHETLAND Is.

ORKNEY Is.

CAITH-NESS

SUTHER-LAND

ROSS & CROMARTY

INVERNESS

NAIRN

MORAY

BANFF

ABER-DEEN

KINCARDINE

ANGUS

PERTH

FIFE

E. LOTHIAN

MID-LOTHIAN

BER-WICK

RENFREW

AYR

DUMFRIES

KIRK-CUDBRIGHT

WIG-TOWN

NORTH-UMBERLAND

SKYE

INNER

HEBRIDES

ARGYLL

OUTER HEBRIDES

DONEGAL

LONDONDERRY

CPSIA information can be obtained at www.ICGtesting.com
Printed in the USA
BVOW06s0323070516

447015BV00006B/17/P